INTERNATIONAL UNION OF PURE AND APPLIED CHEMISTRY

CHEMISTRY INTERNATIONAL

SUPPLEMENT

IUPAC HANDBOOK 1983–1985

PERGAMON PRESS

OXFORD · NEW YORK · TORONTO · SYDNEY · PARIS · FRANKFURT

U.K.	**Pergamon Press Ltd., Headington Hill Hall, Oxford OX3 0BW, England**
U.S.A.	**Pergamon Press Inc., Maxwell House, Fairview Park, Elmsford, New York 10523, U.S.A.**
CANADA	**Pergamon Press Canada Ltd., Suite 104, 150 Consumers Road, Willowdale, Ontario M2J 1P9, Canada**
AUSTRALIA	**Pergamon Press (Aust.) Pty. Ltd., P.O. Box 544, Potts Point, N.S.W. 2011, Australia**
FRANCE	**Pergamon Press SARL, 24 rue des Ecoles, 75240 Paris, Cedex 05, France**
FEDERAL REPUBLIC OF GERMANY	**Pergamon Press GmbH, Hammerweg 6, D-6242 Kronberg-Taunus, Federal Republic of Germany**

First edition 1984

ISBN 0 08 031437 6

Published as a supplement to the journal *Chemistry International,* Volume 1984 and supplied to subscribers as part of their subscription. Also available to non-subscribers

Contents

OFFICERS

(as at 31st March 1984)

President:	Dr. W. G. SCHNEIDER, National Research Council of Canada, 100 Sussex Drive, Ottawa, Ontario K1A 0R6, Canada (TEL: +1 (613) 993 9821. TELEX: 533715 NRC OTT CA)
Vice-President:	Prof. C. N. R. RAO, Solid State and Structural Chemistry Unit, Indian Institute of Science, Malleswaram, Bangalore-560012, India (TEL: 34411. TELEX: 0845-8349 IISC IN)
Secretary General:	Prof. T. S. WEST, Macaulay Institute for Soil Research, Craigiebuckler, Aberdeen AB9 2QJ, UK (TEL: +44 (224) 38611. TELEX: 73458 UNIABN G)
Treasurer:	Prof. A. BJÖRKMAN, Instituttet for Kemiindustri, Danmarks Tekniske Højskole, Bygning 227, DK-2800 Lyngby, Denmark (TEL: +45 (2) 88 41 44. TELEX: 37529 DTHDIA DK)

SECRETARIAT

Bank Court Chambers, 2-3 Pound Way, Cowley Centre, Oxford OX4 3YF, UK
(TEL: +44 (865) 717744. TELEGRAMS: IUPAC OXFORD. TELEX: 83147 ATTN. IUPAC G)

Dr. M. WILLIAMS
Executive Secretary

Miss ANN TROUGHTON
Deputy Executive Secretary

Mr. P. D. GUJRAL
Scientific Publications Secretary

Mr. R. W. FENNELL
Editor of Chemistry International

ORGANIZATION OF IUPAC (1983-1985)

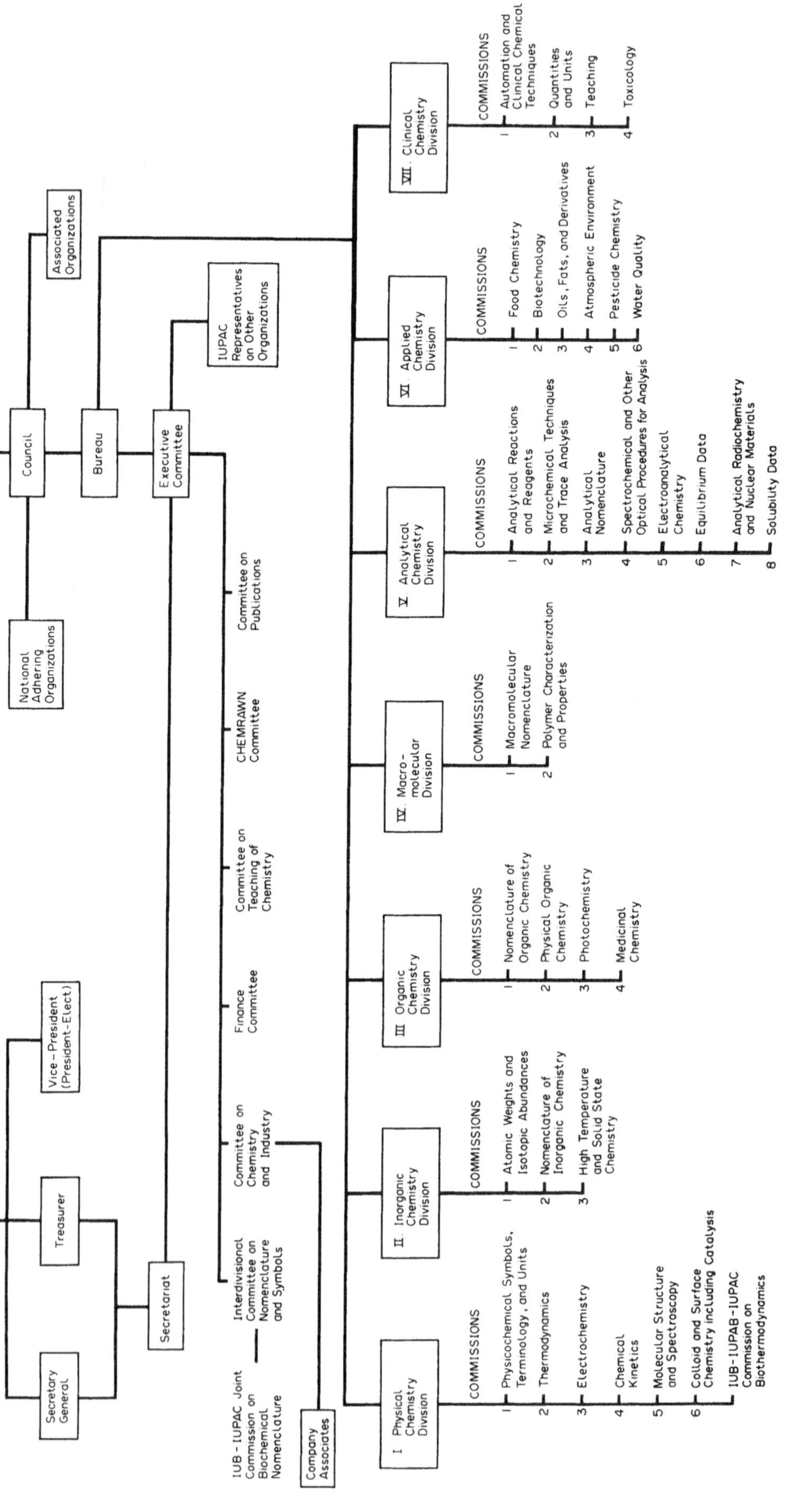

TABLE OF CONTENTS
MEMBERSHIP LISTS

Page

Dates indicate when persons were first elected to their present category of Membership of a Commission. For all other IUPAC bodies they indicate when the Member was first elected. Dates do not necessarily reflect the term of an Officer of an IUPAC body.

MEMBERSHIP LISTS OF IUPAC BODIES 1983 — 1985

NATIONAL ADHERING ORGANIZATIONS

Arab Republic of Egypt	Academy of Scientific Research and Technology, 101 Kasr El-Ainy Street, Cairo, Arab Republic of Egypt
Argentina	Associacíon Química Argentina, Sánchez de Bustamente 1749, Buenos Aires, Argentina (TEL: +54 (1) 8244096/834886)
Australia	Australian Academy of Science, POB 783, Canberra City, ACT 2601, Australia (TEL: +61 (62) 48 6011)
Austria	Gesellschaft Österreichischer Chemiker, Eschenbachgasse 9, A-1010 Wien, Austria (TEL: +43 (222) 574249)
Belgium	Comité National Belge de Chimie, Palais des Académies, 1 rue Ducale, B-1000 Bruxelles, Belgium
Brazil	Associação Brasileira de Química, Caixa Postal 550-CEP 20.043, Av. Rio Branco 156-s/907, Edif. Av. Central, Rio de Janeiro, Brazil (TEL: +55 (21) 262 1837)
Bulgaria	Bulgarian Academy of Sciences, Ul. 1 "7 noemvri", Sofia 1000, Bulgaria (TEL: 729820. TELEX: 22424 BG)
Canada	National Research Council of Canada, Montreal Road, Ottawa, Ontario K1A OR6, Canada (TEL: +1 (613) 993 2024)
China	Chinese Chemical Society, POB 2709, Beijing, China (TEL: 284020. TELEX: 9422 CN)
	Chemical Society located in Taipei, China, POB 609, Taipei, Taiwan
Colombia	Ministerio de Minas y Petróleos, Sub-Dirección de Investigaciones Químicas del Instituto Nacional de Investigaciones Geologico-Mineras, Apartado Aéro 4865, Carrera 30, Numero 51-59, Bogotá, DE, Colombia (TEL: 443330)
Cuba	Academia de Ciencias de la República de Cuba, Capitolio Nacional, La Habana, Cuba
Czechoslovakia	Czechoslovak National Committee for Chemistry, Hradčanské náměstí 12, CS-118 29 Praha 1-Hradčany, Czechoslovakia (TEL: +42 (2) 539074)
Denmark	Det Kongelige Danske Videnskabernes Selskab, Dantes Plads 5, DK-1556 København V, Denmark
Federal Republic of Germany	Deutscher Zentralausschuss für Chemie, Postfach 900440, Carl-Bosch-Haus, Varrentrappstrasse 40-42, D-6000 Frankfurt/Main 90, Federal Republic of Germany (TEL: +49 (611) 79171. TELEX: 412526 GMELI D FÜR GDCH)

Finland	Suomen Kemian Seura, Pohj. Hesperiankatu 3 B 10, SF-00260 Helsinki 26, Finland (TEL: +358 (0) 408022)
France	Comité National Français de la Chimie, 28 rue Saint-Dominique, F-75007 Paris Cédex 07, France (TEL: +33 (1) 705 10 73)
German Democratic Republic	Akademie der Wissenschaften der DDR, Otto-Nuschke-Strasse 22-23, DDR-108 Berlin, German Democratic Republic (TEL: +37 (2) 200481)
Greece	Association of Greek Chemists, 27 Kaningos Street, Athens 10 682, Greece (TEL: +30 (1) 3633597/3621524)
Hungary	Hungarian Academy of Sciences, POB 6, Münnich F.u. 7, H-1361 Budapest, Hungary (TEL: +36 (1) 383 386)
India	Indian National Science Academy, Bahadur Shah Zafar Marg, New Delhi-110002, India (TEL: +91 (11) 276730)
Iraq	Iraqi Chemical Society, POB 8011, Damascus Street, Salihiyah, Baghdad, Iraq (TEL: +964 (1) 30313)
Ireland	Royal Irish Academy, 19 Dawson Street, Dublin 2, Ireland (TEL: +0044 (0011) 762570/764222)
Israel	Israel Academy of Sciences and Humanities, POB 4040, Albert Einstein Square, Talbich, Jerusalem 91040, Israel (TEL: +972 (2) 636211)
Italy	Consiglio Nazionale delle Ricerche, Piazzale Aldo Moro 7, I-00100 Roma, Italy (TEL: +39 (6) 49931. TELEX: 610076 CNR RM I)
Japan	Science Council of Japan, 22-34 Roppongi 7-chome, Minato-ku, Tokyo, 106 Japan (TEL: +81 (3) 403 6291)
Malaysia	Malaysian Institute of Chemistry, c/o Rubber Research Institute of Malaysia, POB 150, Kuala Lumpur, Malaysia
Netherlands	Koninklijke Nederlandse Chemische Vereniging, Postbus 90613, Burnierstraat 1, 2509 LP Den Haag, Netherlands (TEL: +31 (70) 469406)
New Zealand	Royal Society of New Zealand, Private Bag, 6 Halswell Street, Wellington, New Zealand (TEL: +64 (4) 45516)
Norway	Norsk Kjemisk Selskap, POB 1107, Blindern, Oslo 3, Norway (TEL: +47 (2) 455694/455402)
Peru	Sociedad Química del Perú, Av. Nicolás de Aranibar 696, Santa Beatriz, Apartado 891, Lima 100, Peru (TEL: 723925)
Poland	Polska Akademia Nauk, Pałac Kultury i Nauki, Skrytka Pocztowa 24, PL-00 901 Warszawa, Poland (TEL: +48 (22) 200211)
Portugal	Sociedade Portuguesa de Química, Avenida da República 37-4°, P-1000 Lisboa, Portugal (TEL: +351 (1) 734637)

Republic of Korea	Korean Chemical Society, 35, 5-KA Anam Dong, Sungbuk-ku, Seoul 132, Republic of Korea (TEL: 945447)
Republic of South Africa	Council for Scientific and Industrial Research, POB 395, Pretoria 0001, Republic of South Africa (TEL: +27 (12) 749111. TELEX: 3-630 SA)
Romania	Academia Republicii Socialiste România, Calea Victoriei 125, București, Romania
Spain	Consejo Superior de Investigaciones Científicas, Serrano 117, Madrid 6, Spain (TEL: +34 (1) 2619800. TELEX: 42182 E)
Sweden	Svenska Nationalkommittén för Kemi, Upplandsgatan 6 A, 1 tr, S-111 23 Stockholm, Sweden (TEL: +46 (8) 115260/115280)
Switzerland	Schweizerisches Komitee für Chemie, c/o Organisch-Chemisches Institut der Universität Zürich, Winterthurerstrasse 190, CH-8057 Zürich, Switzerland (TEL: +41 (1) 229 42 31)
Syrian Arab Republic	Supreme Council of Sciences, POB 4762, Damascus, Syrian Arab Republic (TEL: 339754)
Turkey	Türkiye Kimya Derneği, Posta Kutusu 829, Harbiye, Halâskâr Gazi Caddesi 53, D 8 Uzay Apartmani, Istanbul, Turkey (TEL: 1407331)
Union of Soviet Socialist Republics	Academy of Sciences of USSR, Leninskii Prospekt 14, Moscow V-71, USSR (TEL: +7 (095) 232 2910. TELEX: 7564 ANS SU)
United Kingdom	Royal Society, 6 Carlton House Terrace, London SW1Y 5AG, UK (TEL: +44 (1) 839 5561. TELEX: 917876 ROYSOC G)
United States of America	National Academy of Sciences, 2101 Constitution Avenue NW, Washington, DC 20418, USA (TEL: +1 (202) 334 2156. TELEX: TWX 710-822-9589)
Venezuela	Instituto Venezolano de Investigaciones Científicas, Centro de Química, Apartado 1827, Caracas 1010 A, Venezuela (TEL: +58 (2) 6811188/691951-59. TELEX: 21338 VE)
Yugoslavia	Unija Hemijskih Društava Jugoslavije, POB 462, Karnegijeva 4, YU-11000 Beograd, Yugoslavia (TEL: +38 (11) 328171)

OBSERVER COUNTRIES

Chile	Sociedad Chilena de Química, Casilla 2613, Concepción, Chile
Hong Kong	Hong Kong Chemical Society, c/o Department of Chemistry, University of Hong Kong, Pokfulam Road, Hong Kong
Malawi	Chemical Society of Malawi, c/o Malawi Bureau of Standards, POB 946, Blantyre, Malawi
Nepal	Nepal Chemical Society, c/o Department of Chemistry, Tribhuwan University, Kiritipur Campus, Kathmandu, Nepal
Pakistan	Chemical Society of Pakistan, c/o Pakistan Science Foundation, PO Box 1121, Islamabad, Pakistan (TEL: 3847)
Singapore	Singapore National Institute of Chemistry, c/o National University of Singapore, Kent Ridge, Singapore 0511 (TEL: 7756666. TELEX: 33943 UNISPO RS)
Sri Lanka	Institute of Chemistry, 120/10, Vidya Mandiraya, Vidya Mawatha, Colombo 7, Sri Lanka
Thailand	Chemical Society of Thailand, c/o Faculty of Science, Mahidol University, Rama VI Road, Bangkok, Thailand

ASSOCIATED ORGANIZATIONS

Association of Editors of European Chemistry Journals	Dr. M. V. KISAKÜREK (Secretary EdEuChem), c/o Helvetica Chimica Acta, Faculté des Sciences, CH-1700 Fribourg, Switzerland
Association of Official Analytical Chemists	Dr. D. B. MacLEAN (Executive Director AOAC), Association of Official Analytical Chemists, Suite 210, 1111 North Nineteenth Street, Arlington, Virginia 22209, USA (TEL: +1 (703) 522 3032)
European Federation of Biotechnology	Prof. D. BEHRENS (EFB General Secretariat), DECHEMA, Postfach 970146, Theodor-Heuss-Allee 25, D-6000 Frankfurt/Main 97, Federal Republic of Germany (TEL: +49 (611) 7564235). TELEX: 0412490 D)
European Federation of Chemical Engineering	Prof. D. BEHRENS (EFCE General Secretariat), DECHEMA, Postfach 970146, Theodor-Heuss-Allee 25, D-6000 Frankfurt/Main 97, Federal Republic of Germany (TEL: +49 (611) 7564235). TELEX: 0412490 D)
European Federation of Corrosion	Prof. D. BEHRENS (EFC General Secretariat), DECHEMA, Postfach 970146, Theodor-Heuss-Allee 25, D-6000 Frankfurt/Main 97, Federal Republic of Germany (TEL: +49 (611) 7564235). TELEX: 0412490 D)
European Photochemistry Association	Prof. F. C. DE SCHRYVER (Chairman EPA), Afdeling Organische Scheikunde, Katholieke Universiteit te Leuven, Celestijnenlaan 200 F, B-3030 Heverlee, Belgium (TEL: +32 (16) 20 06 56. TELEX: 23674 KULEUV B)
Federation of Asian Chemical Societies	Prof. N. S. AL-NIAIMI (Secretary General FACS), Federation of Asian Chemical Societies, c/o Iraqi Chemical Society, POB 8011, Damascus Street, Salihiyah, Baghdad, Iraq
Federation of European Chemical Societies	Dr. W. FRITSCHE (FECS Secretariat), Gesellschaft Deutscher Chemiker, Carl-Bosch-Haus, Postfach 900440, Varrentrappstrasse 40-42, D-6000 Frankfurt/Main 90, Federal Republic of Germany (TEL: +49 (611) 791 7320. TELEX: 41-2526 GMELI D FÜR GDCH)
	Dr. A. JAKABOS (FECS Secretariat), Magyar Kémikusok Egyesülete, Anker Köz 1, H-1061 Budapest, Hungary (TEL: +36 (1) 427 343)
Groupement pour l'Avancement des Méthodes Spectroscopiques et Physico-Chimiques d'Analyse	Dr. P. CROISSANT (Director GAMS), GAMS, 88 boulevard Malesherbes, F-75008 Paris, France (TEL: +33 (1) 563 93 04)
International Association for Advancement of High Pressure Science and Technology	Prof. N. J. TRAPPENIERS (President AIRAPT), Van der Waals-Laboratorium, Universiteit van Amsterdam, Valckenierstraat 67, 1018 XE Amsterdam, Netherlands (TEL: +31 (20) 522911)
International Association for Cereal Chemistry	Dr. H. GLATTES (Secretary General ICC), ICC General Sekretariat, Schmidgasse 3-7, A-2320 Schwechat, Austria (TEL: 777202. TELEX: 133316 A)

International Association of Geochemistry and Cosmochemistry	Dr. SARAH DEUTSCH (Secretary IAGC), Laboratoires de Minéralogie et de Pétrologie, Université Libre de Bruxelles, 50 avenue F. D. Roosevelt, B-1050 Bruxelles, Belgium (TEL: +32 (2) 649 00 30)
International Association on Water Pollution Research and Control	Mr. A. MILBURN (Executive Director IAWPRC), Alliance House, 29-30 High Holborn, London WC1V 6BA, UK (TEL: +44 (1) 404 4083/405 4552. TELEX: 888941 LCCI ATTN. IAWPR G)
International Carbohydrate Organization	Prof. G. N. RICHARDS (Secretary ICO), Department of Chemistry and Biochemistry, James Cook University of North Queensland, Post Office, James Cook University, Queensland 4811, Australia (TEL: Douglas 814343. TELEX: AA47009)
International Commission on Applications of the Mössbauer Effect	Prof. G. M. KALVIUS (Executive Secretary ICAME), Physik-Department E15, Fachbereich Physik, Technische Universität München, James Franck Strasse, D-8046 Garching bei München, Federal Republic of Germany (TEL: +49 (89) 32092501. TELEX: 0522854 D)
International Committee for Science of Photography	Prof. H. PIETSCH (Secretary ICSP), Akademie der Wissenschaften der DDR, Haus 10.2, Rudower Chaussee, DDR-1199 Berlin, German Democratic Republic (TEL: +37 (2) 6702841)
International Committee on Rheology	Prof. N. W. TSCHOEGL (Secretary ICR), Division of Chemistry and Chemical Engineering, California Institute of Technology, Pasadena, California 91125, USA
International Confederation for Thermal Analysis	Prof. F. YARIV (Secretary ICTA), Department of Inorganic and Analytical Chemistry, Hebrew University of Jerusalem, Jerusalem 91904, Israel (TEL: +972 (2) 585214. TELEX: 25391 HU IL)
International Conferences on Coordination Chemistry	Prof. S. KIRSCHNER (Permanent Secretary ICCC), Department of Chemistry, Wayne State University, Detroit, Michigan 48202, USA (TEL: +1 (313) 577 2571)
International Conferences on Organometallic Chemistry	Prof. E. ABEL (Permanent International Secretary ICOMC), Department of Chemistry, University of Exeter, Stocker Road, Exeter EX4 4QD, UK (TEL: +44 (392) 77911)
International Congress of Catalysis	Prof. J. H. BLOCK (Secretary ICC), Abteilung Grenzflächenreaktionen, Fritz-Haber-Institut der Max-Planck-Gesellschaft, Faradayweg 4-6, D-1000 Berlin 33, Federal Republic of Gemany (TEL: +49 (30) 8305411. TELEX: 185676 FHIMP D)
International Organization for Biotechnology and Bioengineering	Ing. C. ROLZ (Chairman IOBB), c/o ICAITI, POB 1552, Guatemala City, Guatemala
International Federation of Clinical Chemistry	Dr. J. G. HILL (Secretary IFCC), Department of Clinical Chemistry, Hospital for Sick Children, 555 University Avenue, Toronto, Ontario M5G 1X8, Canada (TEL: +1 (416) 597 1500. TELEX: 06-22335 CA)
International Organization of Crystal Growth	Prof. M. SCHIEBER (Secretary IOCG), School of Applied Science and Technology, Hebrew University of Jerusalem, Givat Ram, Jerusalem 91904, Israel (TELEX: 25391 HU IL)

International Society of Electrochemistry	Prof. K. V. KORDESCH (Secretary General ISE), Technische Universität Graz, Stremayrgasse 16, A-8010 Graz, Austria
International Society of Heterocyclic Chemistry	Prof. T. KAPPE (Secretary General ISHC), Institut für Organische Chemie der Universität Graz, Heinrichstrasse 28, A-8010 Graz, Austria
International Society of Magnetic Resonance	Prof. D. FIAT (Chairman ISMAR), Medical Center, University of Illinois, POB 6998, Chicago, Illinois 60680, USA (TEL: +1 (312) 996 7609)
International Zeolite Association	Dr. G. T. KERR (President IZA), Mobil Research and Development Corpn., Central Research Division, POB 1025, Princeton, New Jersey 08540, USA
Union of Arab Chemists	Dr. M. H. HNOOSH (Secretary General UAC), Union of Arab Chemists, POB 8011, Damascus Street, Salihiyah, Baghdad, Iraq (TEL: +964 (1) 30313)

COMPANY ASSOCIATES

Arab Republic of Egypt

El-Nasr Pharmaceutical Chemical Co.
NODCAR - National Organization for Drug Control & Research

Belgium

Agfa-Gevaert NV
Janssen Pharmaceutica
Solvay et Cie SA

China

Chinese Petroleum Corpn.
Taiwan Fertilizer Co. Ltd.
Union Chemical Laboratories
USI Far East Corpn.

Czechoslovakia

IRAPA - Vývojový a Racionalizační Ústav Průmyslu
Výzkumný Ústav Pre Petrochémiu

Denmark

Novo Industri A/S
F. L. Smidth & Co.
Superfos A/S
United Breweries Ltd.

Federal Republic of Germany

BASF AG
Bayer AG
C. H. Boehringer Sohn
Chemische Werke Hüls AG
Degussa AG
Dynamit Nobel AG
Th. Goldschmidt AG
Henkel KGaA
Hoechst AG
E. Merck AG
Schering AG
Süd-Chemie AG
Dr. Karl Thomae GmbH
Wacker-Chemie GmbH

Finland

Kemian Keskusliitto

France

ATO-Chimie SA
CECA SA
Cie Française de Raffinage
Hoechst France
Kodak-Pathé SA
L'Oréal SA
Péchiney Ugine Kuhlmann SA
Rhône-Poulenc SA
Roussel Uclaf SA

Greece

General State Chemical Laboratories

India

Indian Petrochemicals Corpn. Ltd.

Ireland

Loctite (Ireland) Ltd.
Penn Chemicals BV
Warner-Lambert (Ireland) Ltd.

Israel

Koor Chemicals Ltd.

Italy

Gruppo Lepetit SpA
Montedison SpA

Japan

Ajinomoto Co. Ltd.
Asahi Chemical Industry Co. Ltd.
Asahi Glass Co. Ltd.
Banyu Pharmaceutical Co. Ltd.
Bridgestone Tire Co. Ltd.
Chugai Pharmaceutical Co. Ltd.
Daicel Chemical Industries Ltd.
Daiichi Seiyaku Co. Ltd.
Daikin Industries Ltd.
Dainippon Pharmaceutical Co. Ltd.
Dainiseikosha Co. Ltd.
Eisai Co. Ltd.
Fuji Photo Film Co. Ltd.
Fujisawa Pharmaceutical Co. Ltd.
Hitachi Ltd.
Ihara Chemical Industry Co. Ltd.
Japan Spectroscopic Co.
JGC Corpn.
Kao Corpn.
Konishiroku Photo Ind. Co. Ltd.
Kumiai Chemical Industry Co. Ltd.
Kuraray Co. Ltd.
Kureha Chemical Industry Co. Ltd.
Kyowa Hakko Kogyo Co. Ltd.
Meiji Seika Kaisha Ltd.
Mitsubishi Chemical Industries Ltd.
Mitsubishi Gas Chemical Co. Inc.
Mitsubishi Petrochemical Co. Ltd.
Mitsubishi Rayon Co. Ltd.
Mitsui Toatsu Chemicals Inc.
Nihon Nohyaku Co. Ltd.
Nippon Carbon Co. Ltd.
Nippon Electric Co. Ltd.
Nippon Mining Co.
Nippon Paint Co. Ltd.
Nippon Steel Chemical Co.
Nippon Steel Corpn.
Nippon Zeon KK Gijutsubu
Nissan Chemical Industries Ltd.

Japan (continued)

Oji Paper Co. Ltd.
Otsuka Pharmaceutical Co. Ltd.
Rigaku Corpn.
Sagami Chuo Kagaku Kenkyusho
Sankyo Co. Ltd.
Sekisui Chemical Co. Ltd.
Shimadzu Corpn.
Shin-etsu Chemical Co. Ltd.
Shionogi & Co. Ltd.
Shiseido Co. Ltd.
Showa Denko KK
Sumitomo Chemical Co. Ltd.
Sumitomo Electric Industries Ltd.
Sumitomo Metal Mining Co. Ltd.
Suntory Ltd.
Taiho Pharmaceutical Co. Ltd.
Taisho Pharmaceutical Co. Ltd.
Takasago Perfumery Co. Ltd.
Takeda Chemical Industries Ltd.
Tama Chemical Co.
Tanabe Seiyaku Co. Ltd.
Teijin Ltd.
Tokuyama Soda Co. Ltd.
Toray Industries Inc.
Yoshida Foundation for Science and Technology

Netherlands

AKZO NV
Dow Chemical (Nederland) BV
DSM Research
Duphar BV
Gist-Brocades NV
Shell Research BV
Philips Research Laboratories
Unilever Research Laboratories

New Zealand

Ivon-Watkins-Dow Ltd.

Norway

Dyno Industrier A/S

Republic of South Africa

African Explosives and Chemical Industries Ltd.
SASOL
Sentrachem Ltd.

Spain

Aiscondel SA

Sweden

Kema Nobel AB

Switzerland

CIBA-GEIGY AG
Firmenich & Co. AG
F. Hoffmann-La Roche & Co. AG
Lonza AG
Nestlé Alimentana AG
Sandoz AG

United Kingdom

Beecham Pharmaceuticals PLC
Blue Circle Industries PLC
British Petroleum Co. Ltd.
Burmah Castrol Co. Ltd.
Cole Group PLC
Distillers Co. Ltd.
Esso Petroleum Co. Ltd.
Glaxo Holdings PLC
Imperial Chemical Industries PLC
Laporte Industries Ltd.
Monsanto PLC
Pfizer PLC
Roche Products Ltd.
RTZ Services Ltd.
Shell UK Ltd.
Unilever PLC.
Wellcome Foundation Ltd.

United States of America

Abbott Laboratories
Allied Corpn.
American Cyanamid Co.
American Hoechst Corpn.
Amoco Chemicals Corpn.
Ashland Chemical Co.
BASF Wyandotte Corpn.
Celanese Research Co.
CIBA-GEIGY Corpn.
Clorax Co.
Corning Glass Works
Dow Chemical Co.
Eastman Kodak Co.
Exxon Research and Engineering Co.
FMC Corpn.
Ford Motor Co.
General Electric Co.
B. F. Goodrich Co.
W. R. Grace & Co.
Gulf Science and Technology Co.
Hercules Inc.
Hoffmann-La Roche Inc.
ICI Americas Inc.
3M Co.
McNeil Pharmaceutical
Merck, Sharp and Dohme Research Laboratories
Mobil Research & Development Corpn.
Monsanto Co.
Olin Corpn.
Pfizer Inc.
Philip Morris USA
Phillips Petroleum Co.
Polaroid Corpn.
E. I. du Pont de Nemours Inc.
Proctor and Gamble Co.
A. H. Robins Co.
Rohm and Haas Co.
Schering Corpn.
Shell Development Co.
Smithkline Beckman Corpn.
Standard Oil Co.
Stauffer Chemical Co.
Syntex Corpn.
Tennessee Eastman Co.
Universal Oil Products Inc.
Upjohn Co.
Warner-Lambert/Park-Davis Pharmaceutical Research Co.
Westinghouse Electric Corpn.
Xerox Corpn.

EXECUTIVE COMMITTEE AND BUREAU STANDING COMMITTEES

EXECUTIVE COMMITTEE

Dr. W. G. Schneider (1981-1985)
President
Prof. C. N. R. Rao (1983-1985)
Vice-President
Prof. T. S. West (1983-1987)
Secretary General
Prof. A. Björkman (1984-1987)
Treasurer
Prof. S. Nagakura (1979-1985)
Past-President

Elected Members

Prof. Y. Jeannin (1983-1987)
Prof. J. Michalski (1979-1985)
Prof. C. G. Overberger (1977-1985)

BUREAU

Dr. W. G. Schneider (1981-1985)
President
Prof. C. N. R. Rao (1977-1985)
Vice-President
Prof. T. S. West (1983-1987)
Secretary General
Prof. A. Björkman (1984-1987)
Treasurer
Prof. S. Nagakura (1975-1985)
Past-President

Elected Members

Dr. A. J. Arvía (1983-1987)
Argentina
Dr. J. W. Barrett (1977-1985)
United Kingdom
Prof. A. D. Campbell (1981-1985)
New Zealand
Prof. Y. Jeannin (1979-1987)
France
Prof. A. Kjaer (1977-1985)
Denmark
Prof. V. A. Koptyug (1979-1987)
Union of Soviet Socialist Republics
Prof. J. Michalski (1977-1985)
Poland
Prof. C. G. Overberger (1977-1985)
United States of America
Prof. D. B. Tonks (1979-1987)
Canada
Prof. A. A. Vlček (1979-1987)
Czechoslovakia
Prof. H. Gg. Wagner (1979-1987)
Federal Republic of Germany

Division Presidents

Dr. D. R. Lide, Jr. (1983-1987)
Physical Chemistry Division
Dr. Mary L. Good (1981-1985)
Inorganic Chemistry Division
Prof. G. Modena (1983-1985)
Organic Chemistry Division
Prof. C. H. Bamford (1981-1985)
Macromolecular Division
Prof. F. Pellerin (1981-1985)
Analytical Chemistry Division
Dr. H. Frehse (1981-1985)
Applied Chemistry Division
Prof. F. W. Sunderman, Jr. (1981-1985)
Clinical Chemistry Division

CHEMRAWN COMMITTEE
(Established 1976)

Dr. B. W. Rossiter (1976-1984)
Chairman
Dr. H. Egan (1976-1984)
Secretary

Members

Prof. J. B. Donnet (1982-1986)
Prof. C. G. Overberger (1976-1984)
Dr. R. Pariser (1982-1986)
Dr. G. S. Sidhu (1983-1987)
Dr. M. Uchida (1979-1987)
Prof. G. E. Zaikov (1976-1984)

Associate Member

Dr. R. H. Marchessault (1982-1986)

(For Standing Orders, see page 39)

COMMITTEE ON CHEMISTRY AND INDUSTRY (COCI)
(Established 1977)

Prof. K. H. Büchel (1978-1986)
Chairman
Dr. G. Jäger (1979-1987)
Secretary

Members

Dr. E. Blair (1983-1987)
Dr. A. Brink (1984-1988)
Dr. R. Degain (1982-1986)
Dr. D. A. A. Fagandini (1983-1987)
Dr. P. L. C. Hao (1982-1986)
Dr. V. Macho (1979-1987)
Dr. W. J. Mijs (1983-1987)
Ir. R. Mormont (1978-1986)
Dr. P. Rhyner (1978-1986)
Dr. A. El-Sawi Shoukry (1979-1987)
Prof. D. Sianesi (1978-1986)
Doc. K. Skagius (1979-1987)
Dr. K. Takeda (1981-1985)

(For Standing Orders, see pages 39/40)

COMMITTEE ON PUBLICATIONS
(Established in its present form 1969)

Dr. D. P. den Os (1978-1986)
Chairman

Members

Dr. S. Maričič (1979-1987)
Mr. T. H. G. Michael (1978-1986)
Dr. G. B. Munsch (1979-1987)
Dr. D. A. O'Sullivan (1981-1989)
Dr. I. A. Williams (1978-1986)

(For Standing Orders, see page 40)

PURE AND APPLIED CHEMISTRY EDITORIAL ADVISORY BOARD

Prof. D. H. Whiffen
Physical Chemistry Division
Prof. E. Fluck
Inorganic Chemistry Division
Prof. N. J. Leonard
Organic Chemistry Division
Prof. M. Mandel
Macromolecular Division
Prof. F. Pellerin
Analytical Chemistry Division
Dr. L. E. Coles
Applied Chemistry Division
Prof. W. C. Purdy
Clinical Chemistry Division

CHEMISTRY INTERNATIONAL EDITORIAL ADVISORY BOARD

Prof. T. S. West (1983-1987)
Secretary General (Chairman)
Dr. H. Egan (1983-1985)
CHEMRAWN Committee
Dr. R. Degain (1983-1985)
Committee on Chemistry and Industry
Dr. D. P. den Os (1983-1985)
Committee on Publications
Prof. L. Paoloni (1983-1985)
Committee on Teaching of Chemistry
Dr. K. L. Loening (1983-1985)
Interdivisional Committee on Nomenclature and Symbols
Prof. K. S. W. Sing (1983-1985)
Physical Chemistry Division
Prof. E. Fluck (1983-1985)
Inorganic Chemistry Division
Prof. N. J. Leonard (1983-1985)
Organic Chemistry Division
Prof. M. Mandel (1983-1985)
Macromolecular Division
Prof. G. den Boef (1983-1985)
Analytical Chemistry Division
Dr. L. E. Coles (1983-1985)
Applied Chemistry Division
Prof. D. B. Tonks (1983-1985)
Clinical Chemistry Division
Mr. R. W. Fennell
Editor Chemistry International
(Secretary)

(For Standing Orders, see pages 40/41)

COMMITTEE ON TEACHING OF CHEMISTRY (CTC)
(Established 1963)

Prof. D. J. Waddington (1977-1985)
Chairman
Prof. M. Chastrette (1981-1989)
Secretary

Members

Prof. Marjorie Gardner (1981-1989)
Prof. E. Giesbrecht (1981-1989)
Prof. Aleksandra Kornhauser (1981-1989)
Prof. D. Samuel (1981-1989)
Prof. J. T. Shimozawa (1983-1991)
Prof. Evdokiya M. Sokolovskaya (1981-1989)

National Representatives

Prof. A. I. Khodair (1981-1985)
Arab Republic of Egypt
Prof. A. H. Guerrero (1971-1985)
Argentina
Dr. R. B. Bucat (1983-1985)
Australia
Dr. Edith Jarisch (1983-1985)
Austria
Prof. G. Geuskens (1978-1985)
Belgium
Dr. R. Isuyama (1983-1985)
Brazil
Prof. D. L. Lazarov (1983-1985)
Bulgaria
Prof. B. T. Newbold (1970-1985)
Canada
Prof. S.-M. Wang (1983-1985)
Chemical Society located in Taipei, China
Prof. T. Hua (1983-1985)
Chinese Chemical Society
Prof. R. Acevedo (1983-1985)
Cuba
Prof. J. Horák (1983-1985)
Czechoslovakia
Prof. E. Rancke-Madsen (1970-1985)
Denmark
Dr. Ursula Hofacker (1975-1985)
Federal Republic of Germany
Prof. H. Saarinen (1983-1985)
Finland
Prof. J. L. Rivail (1982-1985)
France
Prof. H. Bremer (1975-1985)
German Democratic Republic
Prof. Irene Dilaris (1976-1985)
Greece
Prof. C. Szántay (1977-1985)
Hungary
Prof. B. Venkataraman (1983-1985)
India

Mr. P. A. Start (1974-1985)
Ireland
Prof. A. Shani (1983-1985)
Israel
Prof. L. Paoloni (1979-1985)
Italy
Prof. Su Eng Loke (1982-1985)
Malaysia
Dr. Clarisse Habraken (1978-1985)
Netherlands
Mr. T. R. Hitchings (1969-1985)
New Zealand
Mrs. Vivi Ringnes (1980-1985)
Norway
Prof. I. Z. Siemion (1983-1985)
Poland
Dr. Mariana P. B. A. Pereira (1979-1985)
Portugal
Prof. G. C. Gerrans (1977-1985)
Republic of South Africa
Prof. C. I. Simionescu (1970-1985)
Romania
Prof. J. Morcillo Rubio (1978-1985)
Spain
Prof. N. G. Vannerberg (1976-1985)
Sweden
Prof. G. H. Wagnière (1978-1985)
Switzerland
Prof. W. Hakki (1979-1985)
Syrian Arab Republic
Prof. C. B. Şenvar (1983-1985)
Turkey
Prof. C. C. Addison (1979-1985)
United Kingdom
Prof. J. J. Lagowski (1983-1985)
United States of America
Dr. L. Cortês (1971-1985)
Venezuela

(For Standing Orders, see page 41)

FINANCE COMMITTEE (FC)
(Established in its present form 1963)

Dr. H. F. Wilson (1979-1989)
Chairman

Members

Dr. P. J. Agius (1977-1985)
Dr. W. Fritsche (1981-1989)
Dr. A. Hartmann (1979-1987)
Dr. T. Miki (1983-1991)
Prof. G. E. Zaikov (1983-1985)

Ex officio Member

Prof. A. Björkman (1984-1987)

(For Standing Orders, see pages 41/42)

INTERDIVISIONAL COMMITTEE ON NOMENCLATURE AND SYMBOLS (IDCNS)
(Established in its present form 1975)

Dr. K. L. Loening (1976-1984)
Chairman
Dr. R. B. Fox (1979-1987)
Secretary

Members

Prof. D. H. Whiffen (1975-1985)
Physical Chemistry Division
Dr. T. D. Coyle (1981-1985)
Inorganic Chemistry Division
Prof. V. Gold (1979-1987)
Organic Chemistry Division
Prof. A. D. Jenkins (1979-1987)
Macromolecular Division
Prof. M. Grasserbauer (1983-1987)
Analytical Chemistry Division
Mr. S. G. Luxon (1983-1987)
Applied Chemistry Division
Dr. R. Zender (1983-1987)
Clinical Chemistry Division

Associate Members

Prof. K. Kuchitsu (1983-1987)
Physical Chemistry Division
Prof. J. Reedijk (1981-1985)
Inorganic Chemistry Division
Dr. W. H. Powell (1983-1987)
Organic Chemistry Division
Prof. W. Ring (1979-1987)
Macromolecular Division
Prof. Yu. A. Zolotov (1981-1985)
Analytical Chemistry Division
Dr. H. Egan (1983-1987)
Applied Chemistry Division
Dr. M. Roth (1981-1985)
Clinical Chemistry Division

Bureau International des Poids et Mesures

Dr. T. J. Quinn

International Organization for Standardization

Mr. A. Thor

International Union of Biochemistry

Dr. H. B. F. Dixon

International Union of Nutritional Sciences

Dr. S. R. Ames

International Union of Pure and Applied Physics

Dr. B. W. Petley

(For Standing Orders, see pages 42/43)

IUB-IUPAC JOINT COMMISSION ON BIOCHEMICAL NOMENCLATURE (JCBN)
(Attached to IDCNS. Established in its present form 1977)

Dr. H. B. F. Dixon (IUB)
Chairman
Dr. A. Cornish-Bowden (IUPAC, 1983-1987)
Secretary

Members

Prof. C. Liébecq (IUB)
Dr. K. L. Loening (IUPAC, 1977-1985)
Dr. G. P. Moss (IUPAC, 1978-1986)
Prof. J. Reedijk (IUPAC, 1981-1985)
Prof. S. F. Velick (IUB)
Prof. J. F. G. Vliegenthart (IUB)

Associate Member

Prof. F. W. Sunderman, Jr.
(IUPAC, 1984-1988)

(For Standing Orders, see pages 43/44)

IUB-IUPAB-IUPAC COMMISSION ON BIOTHERMODYNAMICS (CB)
(Attached to Physical Chemistry Division. Established 1975)

Prof. I. Wadsö (IUPAC, 1975-1984)
Chairman

Members

Prof. H. Eisenberg (IUPAB)
Prof. M. Klingenberg (IUB)
Prof. W. Pfeil (IUB)

Associate Members

Dr. A. E. Beezer
Prof. J. P. Belaich
Prof. S. J. Gill
Prof. L. G. Hepler
Dr. H.-J. Hinz
Prof. G. Rialdi
Prof. H. Suga

(For Standing Orders, see page 44)

IUPAC REPRESENTATIVES ON OTHER ORGANIZATIONS

ICSU General Committee

Prof. T. S. West (1983-1985)

ICSU Commission on Application of Science to Agriculture, Forestry and Aquaculture (CASAFA)

Prof. J. B. Donnet (1982-1985)

ICSU Committee on Data for Science and Technology (CODATA)

Prof. E. F. Westrum, Jr. (1983-1985)

ICSU Committee on Science and Technology in Developing Countries (COSTED)

Prof. D. J. Waddington (1981-1985)

ICSU Committee on Space Research (COSPAR)

Prof. C. Ponnamperuma (1976-1985)

ICSU Committee on Teaching of Science (CTS)

Prof. D. J. Waddington (1979-1985)

ICSU Scientific Committee on Genetic Experimentation (COGENE)

Dr. R. W. F. Hardy (1981-1985)

ICSU Scientific Committee on Problems of the Environment (SCOPE)

Dr. P. C. Kearney (1982-1985)

Inter-Union Commission on Spectroscopy (IUCS)

Dr. R. N. Jones (1979-1985)
Dr. D. W. Turner (1979-1985)

International Congress on Catalysis

Prof. J. Haber (1979-1985)

International Federation of Clinical Chemistry (IFCC)

Prof. F. W. Sunderman, Jr. (1981-1985)
Prof. N. Montalbetti (1981-1985)

International Organization of Chemical Sciences in Development (IOCD)

Prof. C. N. R. Rao (1983-1985)

ISO/TC 12: Quantities, Symbols, Conversion Factors, and Conversion Tables

Prof. D. H. Whiffen (1977-1985)

IUPAP Commission on Symbols, Units, Nomenclature, Atomic Masses, and Fundamental Constants

Prof. D. H. Whiffen (1976-1985)

International Union of Toxicology (IUTOX)

Prof. F. W. Sunderman, Jr. (1982-1985)

International Committee for Standardization in Haematology (ICSH)

Prof. M. Hjelm (1981-1985)

Commission on World Standards in Clinical Pathology (COWS) of World Association of Societies of Pathology (WASP)

Prof. K. Okuda (1983-1985)

IUPHAR Section on Toxicology

Dr. K. Kühn (1983-1985)

PHYSICAL CHEMISTRY DIVISION (I)

(Established 1949)

DIVISION COMMITTEE

Dr. D. R. Lide, Jr. (1979-1987)
President
Prof. K. I. Zamaraev (1979-1987)
Vice-President
Prof. N. Sheppard (1975-1985)
Past-President
Dr. E. D. Becker (1977-1985)
Secretary

Members

Dr. K. N. Marsh (1981-1985)
Dr. R. Parsons (1981-1985)
Prof. G. M. Schneider (1981-1985)
Prof. K. Tamaru (1983-1987)

Coopted Members

Prof. J. Lu
Dr. S. Veprek
Prof. I. Wadsö

SUBCOMMITTEE ON PHYSICOCHEMICAL MEASUREMENTS AND STANDARDS

Dr. K. N. Marsh
Chairman
Dr. A. J. Head
Secretary

Members

Dr. E. Brunner
Prof. L. Crovini
Mr. G. Girard
Prof. D. V. S. Jain
Dr. E. Juhász
Dr. W. Künzel
Prof. B. Le Neindre
Mr. S. D. Rasberry
Dr. R. Sabbah
Prof. K. Wandelburg
Drs. H. F. van Wijk

SUBCOMMITTEE ON PLASMA CHEMISTRY

Dr. S. Veprek
Chairman
Prof. M. Boulos
Vice-Chairman

Members

Prof. J. Amouroux
Prof. R. Avni
Prof. R. D'Agostino
Dr. J. Dieleman
Dr. H. Drost
Prof. P. Fauchais
Dr. D. L. Flamm
Dr. G. K. Herb
Prof. W. Lindinger
Dr. S. Morita
Prof. E. Pfender
Dr. Z. W. Sternberg
Dr. B. Waldie
Prof. H. Wilhelmi

COMMISSION ON PHYSICOCHEMICAL SYMBOLS, TERMINOLOGY, AND UNITS (I.1)
(Established in its present form 1936)

Titular Members

Prof. K. Kuchitsu (1979-1987)
Chairman
Dr. N. Kallay (1977-1985)
Secretary
Prof. E. T. Denisov (1981-1985)
Prof. K. H. Homann (1979-1987)
Prof. I. M. Mills (1983-1987)
Mr. D. D. Wagman (1981-1985)

Associate Members

Prof. R. A. Alberty (1983-1987)
Dr. I. Ansara (1983-1987)
Dr. E. R. Cohen (1979-1987)
Prof. A. Schuyff (1979-1987)
Prof. D. H. Whiffen (1977-1985)

National Representatives

Prof. M. A. G. Cecchini (1983-1985)
Brazil
Prof. L. Treindl (1981-1985)
Czechoslovakia
Dr. E. Juhász (1983-1985)
Hungary
Dr. S. P. Koh (1983-1985)
Malaysia

COMMISSION ON THERMODYNAMICS (I.2)
(Established in its present form 1977)

Titular Members

Prof. G. M. Schneider (1975-1985)
Chairman
Dr. S. Angus (1977-1985)
Secretary
Prof. R. D. Freeman (1983-1987)
Dr. V. A. Medvedev (1977-1985)

Prof. G. Somsen (1983-1987)
Prof. W. Zielenkiewicz (1977-1985)

Associate Members

Prof. J. M. G. Barthel (1983-1987)
Prof. J. C. G. Calado (1983-1987)
Dr. M. B. Ewing (1981-1985)
Dr. H. V. Kehiaian (1981-1985)
Prof. I. L. Khodakovskii (1983-1987)
Prof. P. A. G. O'Hare (1981-1985)
Dr. Gerd Olofsson (1981-1985)
Prof. J. Pick (1983-1987)
Prof. Margit T. Rätzsch (1981-1985)
Dr. J. Rouquérol (1981-1985)
Prof. H. Suga (1981-1985)

National Representatives

Prof. J. C. Ahluwalia (1983-1985)
India
Prof. I. Michaeli (1983-1985)
Israel
Dr. D. Marchidan (1981-1985)
Romania

SUBCOMMITTEE ON THERMODYNAMIC TABLES

Dr. H. V. Kehiaian
Chairman
Dr. H. J. White, Jr.
Secretary

Members

Dr. S. Angus
Prof. F. Kohler
Dr. S. Malanowski
Dr. K. N. Marsh
Dr. W. A. Wakeham
Prof. K. Watanabe

SUBCOMMITTEE ON TRANSPORT PROPERTIES

Prof. J. Kestin
Chairman
Dr. W. A. Wakeham
Secretary

Members

Dr. S. Angus
Dr. H. J. M. Hanley
Prof. A. Nagashima
Prof. C. A. Nieto de Castro
Prof. H. A. Øye
Prof. K. Stephan

COMMISSION ON ELECTROCHEMISTRY (I.3)
(Established 1951)

Titular Members

Prof. S. Trasatti (1975-1985)
Chairman
Dr. K. Niki (1977-1985)
Secretary
Prof. L. R. Faulkner (1983-1987)
Prof. G. Gritzner (1983-1987)
Prof. D. Landolt (1983-1987)
Dr. R. Memming (1983-1987)

Associate Members

Dr. J. N. Agar (1979-1987)
Dr. A. Bewick (1983-1987)
Prof. E. Budevski (1979-1987)
Prof. K. E. Heusler (1983-1987)
Dr. J. C. Justice (1983-1987)
Prof. J. Koryta (1983-1987)
Prof. V. M. M. Lobo (1983-1987)
Dr. B. Miller (1979-1987)
Dr. R. Parsons (1981-1985)
Prof. O. A. Petrii (1983-1987)
Dr. J. A. Plambeck (1979-1987)
Dr. A. Yamada (1983-1987)

National Representatives

Dr. A. J. Arvía (1980-1985)
Argentina
Prof. T. Rabockay (1983-1985)
Brazil
Prof. B. E. Conway (1979-1985)
Canada
Dr. G. Horányi (1979-1985)
Hungary
Prof. S. K. Rangarajan (1981-1985)
India
Prof. E. Gileadi (1983-1985)
Israel
Prof. J. W. Tomlinson (1973-1985)
New Zealand
Prof. Z. Zembura (1983-1985)
Poland
Dr. R. L. Paul (1983-1985)
Republic of South Africa
Prof. E. Mattsson (1981-1985)
Sweden
Dr. A. K. Covington (1979-1985)
United Kingdom
Prof. D. Dražić (1981-1985)
Yugoslavia

COMMISSION ON CHEMICAL KINETICS (I.4)
(Established 1983)

Members to be appointed

COMMISSION ON MOLECULAR STRUCTURE AND SPECTROSCOPY (I.5)
(Established 1955)

Titular Members

Prof. J. R. Durig (1981-1985)
  Chairman
Dr. H. A. Willis (1983-1987)
  Secretary
Prof. A. M. Bradshaw (1983-1987)
Dr. B. G. Derendjaev (1983-1987)
Prof. S. Forsen (1983-1987)
Dr. J. F. J. Todd (1979-1987)

Associate Members

Prof. R. D. Brown (1983-1987)
Dr. Jeannette G. Grasselli (1977-1985)
Prof. R. K. Harris (1981-1985)
Dr. H. Kon (1983-1987)
Prof. Yu. N. Molin (1983-1987)
Prof. W. B. Person (1981-1985)
Prof. H. Ratajczak (1979-1985)
Prof. C. J. H. Schutte (1981-1985)
Prof. M. Tasumi (1981-1985)

National Representatives

Prof. R. Colin (1981-1985)
  Belgium
Prof. Y. Kawano (1983-1985)
  Brazil
Dr. G. Herzberg (1977-1985)
  Canada
Prof. J. Lu (1981-1985)
  Chinese Chemical Society
Prof. Z. Luz (1983-1985)
  Israel
Prof. S. Ng (1983-1985)
  Malaysia
Prof. Bogusława Jeżowska-Trzebiatowska (1979-1985)
  Poland

SUBCOMMITTEE ON INFRARED AND RAMAN SPECTROSCOPY

Prof. G. Zerbi
  Chairman

Members

Prof. S. Brodersen
Prof. A. R. H. Cole
Prof. M. Delhaye
Prof. L. A. Gribov
Dr. R. N. Jones
Prof. W. Kiefer
Dr. A. L. Smith
Prof. A. H. Zewail

SUBCOMMITTEE ON MASS SPECTROSCOPY

Dr. J. F. J. Todd
  Chairman

Members

Prof. J. H. Beynon
Prof. K. Biemann
Prof. H.-F. Grützmacher
Prof. A. Mandelbaum
Prof. J. Marsel
Prof. E. Roth
Prof. V. I. Tal'rose
Prof. M. Tsuchiya

SUBCOMMITTEE ON NOTATIONS AND CONVENTIONS FOR MOLECULAR SPECTROSCOPY

Prof. C. J. H. Schutte
  Chairman

Members

Dr. P. R. Bunker
Prof. M. A. Elyashévich
Dr. G. Herzberg
Dr. J. T. Hougen
Prof. I. M. Mills
Prof. R. S. Mulliken
Prof. P. M. de Wolff

COMMISSION ON COLLOID AND SURFACE CHEMISTRY INCLUDING CATALYSIS (I.6)
(Established in its present form 1979)

Titular Members

Prof. K. S. W. Sing (1983-1987)
  Chairman
Prof. J. Haber (1977-1985)
  Vice-Chairman
Prof. E. Wolfram (1979-1987)
  Secretary
Prof. J. H. Block (1983-1987)
Dr. P. Stenius (1983-1987)
Dr. Lisbeth Ter-Minassian-Saraga (1983-1987)

Associate Members

Dr. N. V. Churaev (1979-1987)
Dr. P. C. Gravelle (1977-1985)
Prof. J. W. Hightower (1983-1987)
Prof. R. J. Hunter (1979-1987)
Dr. A. Maroto (1983-1987)
Dr. J. Rouquérol (1983-1987)
Prof. P. B. Wells (1983-1987)

National Representatives

Prof. G. F. Froment (1981-1985)
  Belgium
Prof. L. G. Ionescu (1983-1985)
  Brazil

Dr. S. G. Mason (1979-1985)
  Canada
Prof. A. Weiss (1983-1985)
  Federal Republic of Germany
Prof. K. Meyer (1981-1985)
  German Democratic Republic
Prof. L. G. Nagy (1979-1985)
  Hungary
Prof. J. C. Kuriacose (1983-1985)
  India
Prof. A. S. Kertes (1979-1985)
  Israel
Prof. A. Kitahara (1979-1985)
  Japan
Prof. J. Lyklema (1983-1985)
  Netherlands
Dr. J. A. Pajares (1983-1985)
  Spain
Prof. D. H. Everett (1975-1985)
  United Kingdom

SUBCOMMITTEE ON CATALYST CHARACTERIZATION

Prof. J. Haber
  Chairman
Prof. J. H. Block
  Secretary

Members

Prof. G. F. Froment
Dr. P. C. Gravelle
Prof. R. S. Hansen
Prof. R. A. W. Haul
Prof. J. W. Hightower
Dr. J. A. Pajares
Prof. L. Riekert
Prof. M. W. Roberts
Prof. K. S. W. Sing
Prof. K. Tamaru
Prof. P. B. Wells

SUBCOMMITTEE ON CHEMICAL PHYSICS OF SURFACES

Prof. R. S. Hansen
  Chairman

Members

Prof. J. H. Block
Dr. P. C. Gravelle
Prof. J. Haber
Prof. K. Meyer
Prof. M. W. Roberts
Prof. K. Tamaru

INORGANIC CHEMISTRY DIVISION (II)

(Established 1949)

DIVISION COMMITTEE

Dr. Mary L. Good (1979-1985)
President
Prof. Kazuo Saito (1977-1985)
Vice-President
Prof. E. Fluck (1981-1985)
Secretary

Members

Prof. M. T. Beck (1983-1987)
Dr. R. Hagemann (1983-1987)
Dr. J. Hlaváč (1981-1985)
Prof. A. B. P. Lever (1983-1987)
Prof. A. M. Sargeson (1983-1987)

Coopted Members

Prof. N. N. Greenwood
Dr. N. E. Holden
Prof. Y. Jeannin)
Prof. K. L. Komarek) Ex officio
Prof. R. L. Martin)

COMMISSION ON ATOMIC WEIGHTS AND ISOTOPIC ABUNDANCES (II.1)
(Established in its present form 1979)

Titular Members

Prof. R. L. Martin (1975-1985)
Chairman
Dr. J. R. de Laeter (1983-1987)
Secretary
Prof. R. C. Barber (1979-1987)
Dr. I. L. Barnes (1975-1985)
Prof. T. L. Chang (1983-1987)
Dr. T. J. Murphy (1979-1987

Associate Members

Dr. T. B. Coplen (1983-1987)
Prof. P. De Bièvre (1981-1985)
Prof. K. G. Heumann (1983-1987)
Dr. N. E. Holden (1983-1987)
Prof. W. H. Johnson (1981-1985)
Dr. H. S. Peiser (1975-1985)
Dr. C. E. Rees (1983-1987)
Dr. K. J. R. Rosman (1983-1987)
Dr. Masako Shima (1981-1985)

National Representatives

Prof. P. S. Goel (1983-1985)
India
Prof. J. R. Gat (1983-1985)
Israel
Prof. M. Ghazali bin Hj. Abdul Rahman (1983-1985)
Malaysia
Prof. V. I. Goldanskii (1979-1985)
Union of Soviet Socialist Republics
Prof. N. N. Greenwood (1984-1985)
United Kingdom

COMMISSION ON NOMENCLATURE OF INORGANIC CHEMISTRY (II.2)
(Established 1921)

Titular Members

Prof. Y. Jeannin (1971-1985)
Chairman
Prof. D. H. Busch (1979-1987)
Vice-Chairman
Dr. T. D. Coyle (1979-1987)
Secretary
Prof. E. Fluck (1979-1987)
Dr. Piroska Fodor-Csányi (1979-1987)
Dr. G. J. Leigh (1973-1985)
Prof. J. Reedijk (1979-1987)
Dr. E. Samuel (1981-1985)

Associate Members

Dr. M. W. G. de Bolster (1983-1987)
Dr. K. Ch. Buschbeck (1979-1985)
Prof. J. A. Connor (1981-1985)
Mr. E. W. Godly (1979-1987)
Dr. R. Laitinen (1981-1985)
Prof. R. C. Mehrotra (1981-1985)
Prof. J. F. Nixon (1981-1985)
Dr. K. Samuelsson (1981-1985)
Mr. T. E. Sloan (1979-1987)

National Representatives

Prof. L. F. Bertello (1979-1985)
Argentina
Prof. A. M. Sargeson (1980-1985)
Australia
Prof. A. C. Massabni (1983-1985)
Brazil
Dr. M. Zikmund (1978-1985)
Czechoslovakia
Prof. K. Yamasaki (1979-1985)
Japan
Dr. L. Y. Goh (1983-1985)
Malaysia
Dr. D. Purdela (1978-1985)
Romania
Prof. S. Fallab (1984-1985)
Switzerland
Dr. V. Simeon (1981-1985)
Yugoslavia

COMMISSION ON HIGH TEMPERATURE AND SOLID STATE CHEMISTRY (II.3)
(Established in its present form 1981)

Titular Members

Prof. K. L. Komarek (1977-1985)
Chairman
Prof. P. W. Gilles (1977-1985)
Secretary
Dr. Anne-Marie Anthony (1977-1985)
Prof. L. V. Gurvich (1979-1987)
Prof. R. Metselaar (1983-1987)
Prof. G. Petzow (1979-1985)
Dr. M. H. Rand (1979-1987)

Associate Members

Prof. C. B. Alcock (1979-1985)
Prof. J. Corish (1983-1987)
Dr. J.-P. Coutures (1979-1985)
Prof. J. Drowart (1981-1985)
Prof. E. Fitzer (1977-1985)
Dr. J. Hastie (1981-1985)
Dr. M. G. Hocking (1979-1987)
Prof. L. Kihlborg (1983-1987)
Dr. R. W. Ohse (1973-1985)
Prof. R. Sersale (1977-1985)
Prof. W. L. Worrell (1977-1985)
Prof. H. Yanagida (1983-1987)

National Representatives

Dr. M. S. E. El-Sewefy (1980-1985)
Arab Republic of Egypt
Dr. E. R. McCartney (1969-1985)
Australia
Prof. T.-S. Yen (1980-1985)
Chinese Chemical Society
Prof. J. Matoušek (1981-1985)
Czechoslovakia
Dr. J. F. Baumard (1982-1985)
France
Prof. R. Vernerkar (1983-1985)
India
Prof. G. De Maria (1982-1985)
Italy
Dr. M. Badri (1983-1985)
Malaysia
Dr. J. B. Clark (1983-1985)
Republic of South Africa
Prof. A. Magnéli (1969-1985)
Sweden
Prof. G. Bayer (1975-1985)
Switzerland
Prof. G. M. Rosenblatt (1978-1985)
United States of America
Prof. M. M. Ristić (1975-1985)
Yugoslavia

SUBCOMMITTEE ON CHARACTERIZATION AND TERMINOLOGY OF CARBON AND GRAPHITE

Prof E. Fitzer
Chairman

Members

Prof. R. C. Bansal
Prof. H. P. Boehm
Prof. L. Bonnetain
Prof. R. J. Diefendorf
Prof. J. B. Donnet
Dr. W. B. Eatherly
Dr. G. B. Engle
Prof. D. B. Fischbach
Prof. M. Inagaki
Dr. B. McEnaney
Prof. H. Marsh
Dr. R. A. Meyer
Dr. E. G. Morris
Prof. S. Mrozowski
Dr. F. Rozploch
Dr. R. Setton
Prof. E. Stumpp
Prof. T. Tsuzuku
Prof. A. R. Ubbelohde

ORGANIC CHEMISTRY DIVISION (III)

(Established 1949)

DIVISION COMMITTEE

Prof. G. Modena (1975-1985)
President
Prof. J. F. Bunnett (1977-1985)
Vice-President
Prof. J. Mathieu (1977-1985)
Past-President
Dr. Sukh Dev (1981-1987)
Secretary

Members

Prof. N. J. Leonard (1981-1985)
Prof. M. Ōki (1983-1987)
Prof. M. Protiva (1981-1985)
Prof. R. A. Rossi (1983-1987)

Coopted Members

Prof. Irine P. Beletskaya
Prof. S. Gronowitz
Prof. W.-Y. Huang
Dr. W. Koch
Dr. J. T. Marvel
Prof. K. Schaffner

National Representatives

Prof. J. A. Retamar (1980-1985)
Argentina
Prof. J. R. Cannon (1982-1985)
Australia
Prof. G. L'abbe (1981-1985)
Belgium
Prof. Maria A. Kaplan (1983-1985)
Brazil
Prof. W.-Y. Huang (1981-1985)
Chinese Chemical Society
Prof. G. Deák (1979-1985)
Hungary
Prof. G. Mehta (1983-1985)
India
Prof. K. C. Chan (1983-1985)
Malaysia
Dr. J. R. Bull (1983-1985)
Republic of South Africa
Prof. A. T. Balaban (1983-1985)
Romania

COMMISSION ON NOMENCLATURE OF ORGANIC CHEMISTRY (III.1)
(Established 1921)

Titular Members

Dr. G. P. Moss (1977-1985)
Chairman
Prof. D. Hellwinkel (1979-1987)
Vice-Chairman
Dr. W. H. Powell (1979-1987)
Secretary
Prof. O. Achmatowicz (1979-1987)
Dr. A. D. McNaught (1979-1987)
Prof. R. Panico (1981-1985)
Prof. J. C. Richer (1979-1987)
Prof. P. A. S. Smith (1983-1987)

Associate Members

Dr. K. Bláha (1979-1987)
Prof. H. J. T. Bos (1983-1987)
Dr. A. J. Boulton (1981-1985)
Dr. P. M. Giles, Jr. (1983-1987)
Mr. E. W. Godly (1979-1987)
Dr. H. Gutmann (1983-1987)
Dr. K. Hirayama (1983-1987)
Mr. S. P. Klesney (1979-1985)
Prof. N. Lozac'h (1979-1987)
Prof. J. Rigaudy (1981-1985)
Mr. R. Schoenfeld (1981-1985)
Dr. O. Weissbach (1979-1987)

COMMISSION ON PHYSICAL ORGANIC CHEMISTRY (III.2)
(Established 1973)

Titular Members

Prof. V. Gold (1983-1985)
Chairman
Dr. R. A. Y. Jones (1981-1985)
Secretary
Prof. E. A. Halevi (1981-1985)
Prof. G. Illuminati (1977-1985)
Prof. W. P. Jencks (1981-1985)
Prof. O. M. Nefedov (1981-1985)
Prof. K. Schwetlick (1977-1985)

Associate Members

Prof. J. F. Bunnett (1983-1987)
Prof. M. P. Doyle (1979-1987)
Prof. R. D. Guthrie (1977-1985)
Dr. J. S. Littler (1979-1987)
Prof. J. March (1977-1985)
Prof. P. Müller (1981-1985)
Dr. J. R. Penton (1981-1985)
Dr. J. Toullec (1981-1985)

National Representatives

Prof. P. Van Brandt (1983-1985)
Belgium
Prof. E. Humeres (1983-1985)
Brazil
Prof. X. Jiang (1980-1985)
Chinese Chemical Society
Prof. M. Nógrádi 1980-1985)
Hungary

Prof. D. J. McLennan (1982-1985)
  New Zealand
Prof. A. T. Balaban (1981-1985)
  Romania
Prof. P. N. I. Ahlberg (1980-1985)
  Sweden
Prof. M. Lj. Mihailović (1979-1985)
  Yugoslavia

WORKING PARTY ON NOMENCLATURE FOR INSERTION AND EXTRUSION TRANSFORMATIONS

Dr. J. Toullec
  Chairman

Members

Prof. M. P. Doyle
Dr. R. A. Y. Jones
Prof. Annick Panaye
Dr. I. Tőth
Prof. P. Van Brandt

WORKING PARTY ON NOMENCLATURE FOR RING-FORMING AND RING-OPENING TRANSFORMATIONS

Prof. P. Müller
  Chairman

Members

Prof. I. Agranat
Prof. M. P. Doyle
Prof. K. N. Houk
Prof. G. W. Klumpp
Prof. G. A. Kraus

WORKING PARTY ON STEREOCHEMICAL TERMINOLOGY

Prof. V. Gold
  Chairman

Members

Dr. J. E. Blackwood
Prof. D. H. Busch
Prof. G. Illuminati
Prof. J. March
Dr. G. P. Moss
Prof. J. C. Richer
Prof. K. Schwetlick
Prof. P. Sigwalt
Dr. O. Weissbach

COMMISSION ON PHOTOCHEMISTRY (III.3)
(Established in its present form 1975)

Titular Members

Prof. F. C. De Schryver (1975-1985)
  Chairman
Prof. Silvia E. Braslavsky (1983-1987)
  Secretary
Prof. K. N. Houk (1977-1985)
Prof. H. Iwamura (1983-1987)
Prof. J. Michl (1981-1985)
Prof. K. Tokumaru (1979-1987)

Associate Members

Dr. D. F. Eaton (1981-1985)
Dr. A. Heller (1983-1987)
Prof. M. G. Kuzmin (1981-1985)
Prof. C. Sandorfy (1977-1985)
Prof. F. Wilkinson (1981-1985)

National Representatives

Prof. A. M. Osman (1973-1985)
  Arab Republic of Egypt
Prof. O. P. Strausz (1979-1985)
  Canada
Prof. K. Lempert (1980-1985)
  Hungary
Prof. M. V. George (1983-1985)
  India
Prof. E. Fischer (1983-1985)
  Israel
Dr. Ibrahim bin Abdullah (1983-1985)
  Malaysia
Prof. J. W. Verhoeven (1981-1985)
  Netherlands

COMMISSION ON MEDICINAL CHEMISTRY (III.4)
(Under Review)

MACROMOLECULAR DIVISION (IV)

(Established 1967)

DIVISION COMMITTEE

Prof. C. H. Bamford (1975-1985)
President
Prof. T. Saegusa (1979-1985)
Vice-President
Prof. M. Mandel (1977-1985)
Secretary

Members

Dr. P. L. Clegg (1983-1987)
Prof. W. Heitz (1983-1987)
Prof. E. M. Pearce (1983-1987)
Prof. N. A. Platé (1981-1985)
Prof. P. Sigwalt (1983-1987)

Coopted Members

Dr. M. E. Carrega
Prof. G. Challa
Dr. J. Economy
Prof. N. Ise
Prof. M. Kryszewski
Dr. H. Lange
Prof. D. C. Pepper
Prof. B. Philipp
Prof. P. Pino
Prof. A. Silberberg
Prof. C. I. Simionescu

National Representatives

Dr. D. H. Solomon (1984-1985)
Australia
Prof. O. Olaj (1979-1985)
Austria
Prof. E. Goethals (1981-1985)
Belgium
Prof. Eloisa B. Mano (1979-1985)
Brazil
Prof. I. M. Panayotov (1968-1985)
Bulgaria
Prof. L. E. St. Pierre (1979-1985)
Canada
Dr. P. Cefelín (1984-1985)
Czechoslovakia
Prof. J. Kops (1974-1985)
Denmark
Prof. W. Ring (1983-1985)
Federal Republic of Germany
Prof. J. J. Lindberg (1971-1985)
Finland
Prof. G. Reinisch (1982-1985)
German Democratic Republic
Prof. G. Hardy (1975-1985)
Hungary
Prof. D. D. Deshpande (1983-1985)
India
Prof. M. Lewin (1983-1985)
Israel
Prof. M. Farina (1976-1985)
Italy
Prof. A. Nakajima (1981-1985)
Japan
Prof. A. Bantjes (1982-1985)
Netherlands
Dr. I. E. Ruyter (1980-1985)
Norway
Prof. Z. J. Jedlínski (1983-1985)
Poland
Prof. E. W. Neuse (1976-1985)
Republic of South Africa
Dr. E. V. Dobrescu (1981-1985)
Romania
Prof. J. G. Fatou (1982-1985)
Spain
Prof. B. Rånby (1969-1985)
Sweden
Prof. V. A. Kabanov (1983-1985)
Union of Soviet Socialist Republics
Prof. D. H. Richards (1983-1985)
United Kingdom
Dr. B. C. Anderson (1983-1985)
United States of America
Dr. E. Paredes (1981-1985)
Venezuela

Representative of ICR

Dr. A. J. de Vries

Representative of IUPAB

Prof. H. A. Scheraga

Representative of IUPAP

Prof. G. W. Becker

COMMISSION ON MACROMOLECULAR NOMENCLATURE (IV.1)
(Established 1968)

Titular Members

Prof. A. D. Jenkins (1975-1985)
Chairman
Dr. N. M. Bikales (1977-1985)
Secretary
Prof. G. Allegra (1977-1985)
Dr. R. E. Bareiss (1983-1987)
Dr. P. Kratochvíl (1979-1987)
Prof. I. Mita (1979-1987)
Prof. I. M. Papisov (1979-1987)
Prof. U. W. Suter (1981-1985)

Associate Members

Prof. D. Braun (1981-1985)
Prof. J. M. G. Cowie (1983-1987)
Dr. L. G. Donaruma (1977-1985)
Dr. K. L. Loening (1977-1985)
Dr. R. H. Marchessault (1983-1987)
Prof. R. P. Quirk (1983-1987)
Prof. Marguerite Rinaudo (1981-1985)
Prof. P. Sigwalt (1983-1987)
Prof. L.-O. Sundelöf (1981-1985)

National Representatives

Prof. Eloisa B. Mano (1979-1985)
Brazil
Prof. L. Shi (1983-1985)
Chinese Chemical Society
Dr. A. S. Tan (1983-1985)
Malaysia
Dr. G. J. Leary (1982-1985)
New Zealand
Prof. N. A. Platé (1983-1985)
Union of Soviet Socialist Republics

COMMISSION ON POLYMER CHARACTERIZATION AND PROPERTIES (IV.2)
(Established 1973)

Titular Members

Dr. P. L. Clegg (1975-1985)
Chairman
Dr. R. E. Wetton (1977-1985)
Secretary
Prof. L. Dulog (1981-1985)
Dr. G. C. Eastmond (1977-1985)
Prof. G. Hinrichsen (1977-1985)
Dr. H. H. Meyer (1983-1987)
Mr. P. Starck (1983-1987)

Associate Members

Prof. Sir Geoffrey Allen (1979-1985)
Dr. B. C. Anderson (1983-1987)
Dr. J. W. Barrett (1981-1985)
Prof. G. W. Becker (1979-1985)
Prof. V. A. Kabanov (1981-1985)
Prof. J. C. Seferis (1983-1987)
Dr. A. J. de Vries (1977-1985)

National Representatives

Prof. N. A. Ghanem (1981-1985)
Arab Republic of Egypt
Prof. R. Qian (1981-1985)
Chinese Chemical Society
Prof. J. Beniska (1981-1985)
Czechoslovakia
Prof. H. Zimmermann (1983-1985)
German Democratic Republic
Dr. T. Kelen (1979-1985)
Hungary
Dr. I. S. Bhardwaj (1983-1985)
India
Prof. M. Levy (1983-1985)
Israel
Dr. A. Subramaniam (1983-1985)
Malaysia
Dr. J. Majnusz (1983-1985)
Poland

WORKING PARTY ON STRUCTURE AND PROPERTIES OF COMMERCIAL POLYMERS

Dr. H. H. Meyer
Chairman
Dr. D. R. Moore
Secretary

Members

Dr. G. Ajroldi
Prof. R. C. Armstrong
Dr. C. B. Bucknall
Mr. J. M. Cann
Dr. D. Constantin
Mr. H. Coster
Dr. van Dijk
Dr. M. Fleissner
Prof. H.-G. Fritz
Dr. P. H. Geil
Mr. A. Ghijsels
Dr. G. Goldbach
Dr. D. J. Groves
Prof. H. Janeschitz-Kriegl
Mr. P. B. Keating
Dr. H. M. Laun
Prof. A. S. Lodge
Prof. C. Macosko
Prof. J. Meissner
Dr. A. Plochocki
Dr. W. Retting
Dr. U. P. Richter
Dr. G. Schorsch
Dr. G. Schoukens
Prof. J. C. Seferis
Dr. J. M. Starita
Dr. G. Vassilatos
Prof. J. L. White
Prof. H. H. Winter
Mr. J. Young
Prof. H. G. Zachmann

JAPANESE SUB-GROUP

Prof. A. Nakajima
Chairman

Members

Dr. S. Hayakawa
Dr. M. Kato
Dr. Y. Kubouchi
Dr. T. Masuda
Dr. T. Ono
Prof. J. Shimizu
Dr. M. Uchida
Dr. A. Yoshioka

SUB-GROUP ON MOLECULAR THEORY

Members

Prof. R. C. Armstrong
Prof. A. S. Lodge
Prof. J. Meissner
Dr. M. H. Wagner
Prof. H. H. Winter

SUB-GROUP ON THE CHARACTERIZATION AND ESTIMATION OF DEFECTS IN THE MOLECULAR STRUCTURE ON POLYVINYLCHLORIDE

Dr. A. Michel
Chairman

Members

Dr. Allsopp
Mme. V. Bellenger
Prof. D. Braun
Dr. Caraculacu
Dr. E. V. Dobrescu
Dr. Elgert
Dr. N. Fischer
Mme. Gallot
Dr. A. Guyot
Dr. C. J. M. van den Heuvel
Dr. P. Kappler
Dr. T. Kelen
Dr. Kolinsky
Mme. Llauro
Dr. Maddams
Dr. Millan
Dr. T. Nagy
Dr. Palma
Dr. Park
Dr. Q. T. Pham
Dr. Simak
Dr. E. Sörvik
Dr. J. A. J. M. Vincent
Dr. H. O. Wirth
Dr. B. F. Witenhafer

WORKING PARTY ON MOLECULAR CHARACTERIZATION OF COMMERCIAL POLYMERS

Mr. P. Starck
Chairman

Members

Dr. W. Ball
Mr. F. Canè
Dr. C. Cometto
Dr. D. Constantin
Dr. R. De Bruille
Dr. A. Holmström
Mr. W. Holtrup
Mr. L. J. Kulin
Dr. C. Leolini
Mr. N. L. J. Meyerink
Prof. B. Philipp
Dr. D. Roussel
Dr. Th. G. Scholte
Dr. C. Strazielle
Mr. S. Å. Sultan
Dr. M. Unbehend
Dr. J. L. Vidal
Dr. F. P. Warner

WORKING PARTY ON SUPPORTED POLYMER FILMS

Prof. L. Dulog
Chairman
Dr. J. Sickfeld
Secretary

Members

Dr. A. Berge
Dr. U. Biethan
Prof. D. Braun
Dr. M. Chiavarini
Mr. G. Christensen
Prof. D. T. Clark
Dr. G. P. Cunningham
Mr. P. H. Fink-Jensen
Mr. A. Fiorelli
Prof. W. Funke
Dr. A. C. Gerster
Mr. M. A. Glaser
Prof. K. Hamann
Dr. Ch. M. Hansen
Siv.ing. S. Heimdahl
Dr. P. Janssen
Dr. S. M. Kambanis
Dr. H. Kollek
Ir. M. F. Kooistra
Dr. H. J. Kretzschmar
Dr. J. A. W. van Laar
Mr. D. H. W. Lovegrove
Dr. H. J. Luthardt
Prof. R. R. Myers
Mr. H. K. Raaschou Nielsen
Dr. K. M. Oesterle
Dr. L. A. O'Neill
Dr. V. A. Ogarev
Prof. D. Pagani
Dr. P. E. Pierce
Dr. E. V. Schmid
Dr. D. H. Solomon
Dr. H. Spoor
Mr. A. R. H. Tawn
Dr. A. Toth
Mr. A. Toussaint
Dr. O. Vorster
Dr. U. Zorll
Mr. V. Zvonar

SUB-GROUP ON SURFACE PROPERTIES
OF FILMS AND FIBRES

Members

Dr. H. R. Anderson
Prof. C. Batich
Dr. D. J. Carlson
Dr. A. Dilks
Dr. R. Jeffries
Dr. D. Sheasley
Dr. H. Shonhorn
Dr. J. P. Sibilia
Dr. D. M. Soignet
Dr. D. W. Williams

WORKING PARTY ON THERMAL PROPERTIES
OF POLYMERS

Prof. G. Hinrichsen
Chairman

Members

Dr. J. M. Barton
Prof. R. F. Boyer
Dr. J. Chiu
Dr. P. Dantinne
Dr. M. Droescher
Dr. J. H. Flynn
Prof. P. D. Garn
Dr. G. Henze
Dr. K. H. Illers
Prof. H. Kambe
Prof. F. E. Karasz
Prof. A. Kettrup
Prof. J. J. Lindberg
Dr. T. R. Manley
Dr. E. Marti
Dr. V. B. F. Mathot
Mr. J. J. Maurer
Mr. J. M. O'Reilly
Dr. J. Paulik
Dr. J. M. Richardson
Dr. P. V. Ronca
Dr. J. P. Schelz
Dr. Edith A. Turi
Dr. I. Vancsó-Szmercsányi
Dr. J. Wendorff
Prof. B. Wunderlich

**WORKING PARTY ON KINETIC PARAMETERS
FOR FREE-RADICAL POLYMERIZATION**

**Dr. G. C. Eastmond
Chairman**

Members

**Prof. C. H. Bamford
Dr. A. Beresniewicz
Dr. K. Berger
Prof. J. C. Bevington
Prof. M. Kamachi
Prof. G. Meyerhoff
Prof. K. F. O'Driscoll
Prof. P. M. Reilly
Prof. P. Rempp
Prof. S. Russo
Dr. M. Stickler
Dr. Y. Y. Tan
Dr. P. Wittmer**

ANALYTICAL CHEMISTRY DIVISION (V)

(Established 1949)

DIVISION COMMITTEE

Prof. F. Pellerin (1975-1985)
President
Prof. G. H. Nancollas (1981-1985)
Vice-President
Prof. G. den Boef (1983-1987)
Secretary

Members

Prof. J. F. Coetzee (1983-1987)
Prof. H. Freiser (1983-1987)
Prof. G. G. Guilbault (1983-1987)
Prof. Y. Marcus (1983-1987)
Prof. J. Robin (1983-1987)
Prof. G. Tölg (1983-1987)

Coopted Members

Prof. J. Inczédy
Prof. J. W. Lorimer
Prof. A. Mizuike
Prof. B. F. Myasoedov
Prof. J. M. Ottaway

Representative of FECS

Prof. E. Pungor

COMMISSION ON ANALYTICAL REACTIONS AND REAGENTS (V.1)
(Established 1934)

Titular Members

Prof. G. Ackermann (1977-1985)
Chairman
Prof. D. Thorburn Burns (1983-1987)
Secretary
Prof. S. B. Savvin (1979-1987)
Prof. L. Sommer (1983-1987)
Dr. W. I. Stephen (1977-1985)
Prof. E. V. Wänninen (1977-1985)

Associate Members

Prof. G. Duyckaerts (1979-1987)
Prof. A. Hulanicki (1979-1987)
Prof. J. Inczédy (1983-1987)
Prof. F. Ingman (1979-1987)
Prof. D. L. Massart (1981-1985)
Prof. D. E. Ryan (1981-1985)
Prof. P. C. Uden (1981-1985)
Prof. K. Ueno (1981-1985)

National Representatives

Prof. N. Jordanov (1979-1985)
Bulgaria
Prof. J. Garaj (1981-1985)
Czechoslovakia
Dr. W. M. Yong (1983-1985)
Malaysia
Dr. Brit Salbu (1982-1985)
Norway

COMMISSION ON MICROCHEMICAL TECHNIQUES AND TRACE ANALYSIS (V.2)
(Established in its present form 1967)

Titular Members

Dr. B. Griepink (1979-1987)
Chairman
Dr. A. Townshend (1977-1985)
Secretary
Prof. K. Biemann (1983-1987)
Prof. E. Jackwerth (1979-1987)
Dr. A. Lamotte (1983-1987)
Prof. Z. Marczenko (1981-1985)
Prof. Yu. A. Zolotov (1983-1987)

Associate Members

Prof. K. Ballschmitter (1983-1987)
Prof. K. Beyermann (1981-1985)
Prof. R. Dams (1983-1987)
Prof. K. Fuwa (1981-1985)
Prof. M. Grasserbauer (1979-1987)
Dr. W. Gries (1983-1987)
Dr. K. Heinrich (1977-1985)
Prof. A. Mizuike (1981-1985)
Prof. J. M. Ottaway (1977-1985)
Prof. G. Tölg (1983-1987)
Dr. D. E. Wells (1983-1987)
Dr. D. G. Westmoreland (1983-1987)

National Representatives

Prof. R. Gijbels (1984-1985)
Belgium
Prof. H. Thoma (1983-1985)
Brazil
Prof. J. Janák (1977-1985)
Czechoslovakia
Dr. Zsuzsa Horváth (1981-1985)
Hungary
Dr. M. Sankar Das (1983-1985)
India
Dr. R. C. H. Hsia (1983-1985)
Malaysia
Prof. A. D. Campbell (1976-1985)
New Zealand
Dr. Brit Salbu (1983-1985)
Norway

Prof. W. Kemula (1979-1985)
Poland
Prof. A. Cedergren (1981-1985)
Sweden
Dr. Eugenia A. Terent'eva (1984-1985)
Union of Soviet Socialist Republics

WORKING GROUP ON ORGANIC TRACE ANALYSIS

Dr. A. Lamotte
Chairman

Members

Prof. K. Ballschmitter
Prof. K. Beyermann
Prof. K. Biemann
Dr. B. Griepink
Dr. D. E. Wells
Dr. D. G. Westmoreland

WORKING GROUP ON SURFACE ANALYSIS

Prof. M. Grasserbauer
Chairman

Members

Dr. W. Gries
Dr. R. Hagemann
Dr. K. Heinrich
Prof. G. H. Morrison
Dr. Christiane Sénémaud

COMMISSION ON ANALYTICAL NOMENCLATURE (V.3)
(Established in its present form 1959)

Titular Members

Dr. G. Svehla (1975-1985)
Chairman
Prof. S. P. Perone (1977-1985)
Secretary
Prof. C. A. M. G. Cramers (1983-1987)
Prof. R. W. Frei (1979-1987)
Prof. D. Klockow (1981-1985)
Prof. R. E. Van Grieken (1983-1987)

Associate Members

Dr. L. A. Currie (1983-1987)
Dr. L. S. Ettre (1981-1985)
Dr. A. E. Fein (1981-1985)
Prof. H. Freiser (1981-1985)
Prof. G. G. Guilbault (1983-1987)
Dr. W. Horwitz (1981-1985)
Prof. B. R. Kowalski (1981-1985)
Dr. M. A. Leonard (1983-1987)
Dr. D. Leyden (1983-1987)
Prof. R. F. Martin (1981-1985)
Dr. B. Schreiber (1981-1985)
Prof. W. Simon (1983-1987)

National Representatives

Prof. I. Giolito (1983-1985)
Brazil
Prof. W. E. Harris (1980-1985)
Canada
Prof. J. Starý (1981-1985)
Czechoslovakia
Prof. R. H. Rosset (1981-1985)
France
Prof. K. Doerffel (1984-1985)
German Democratic Republic
Prof. E. Grushka (1981-1985)
Israel
Prof. H. M. N. H. Irving (1983-1985)
Republic of South Africa
Dr. D. Jagner (1981-1985)
Sweden

COMMISSION ON SPECTROCHEMICAL AND OTHER OPTICAL PROCEDURES FOR ANALYSIS (V.4)
(Established in its present form 1959)

Titular Members

Dr. A. Strasheim (1975-1985)
Chairman
Prof. K. Laqua (1975-1985)
Vice-Chairman
Dr. L. R. P. Butler (1983-1987)
Secretary
Dr. J. M. Mermet (1983-1987)
Dr. I. Rubeška (1977-1985)
Dr. Christiane Sénémaud (1983-1987)

Associate Members

Prof. C. Th. J. Alkemade (1975-1985)
Prof. H. Ebel (1981-1985)
Prof. G. M. Hieftje (1983-1987)
Prof. G. F. Kirkbright (1981-1985)
Prof. B. V. L'vov (1983-1987)
Prof. R. Manne (1981-1985)
Dr. W. H. Melhuish (1983-1987)
Dr. N. S. Nogar (1983-1987)
Prof. E. Plško (1977-1985)
Prof. J. Robin (1981-1985)
Dr. A. M. Ure (1981-1985)
Prof. M. Zander (1983-1987)

National Representatives

Dr. J. H. Cappacioli (1980-1985)
Argentina
Prof. A. J. Curtius (1983-1985)
Brazil
Prof. K. Zimmer (1979-1985)
Hungary
Dr. S. Shibata (1980-1985)
Japan
Dr. L. Pszonicki (1979-1985)
Poland

COMISSSION ON ELECTROANALYTICAL CHEMISTRY (V.5)
(Established in its present form 1963)

Titular Members

Prof. J. Jordan (1975-1985)
Chairman
Prof. K. Izutsu (1977-1985)
Secretary
Dr. A. K. Covington (1977-1985)
Dr. J. Juillard (1979-1987)
Prof. R. C. Kapoor (1977-1985)
Prof. E. Pungor (1979-1987)

Associate Members

Dr. W. Davison (1981-1985)
Dr. R. A. Durst (1979-1987)
Prof. M. Gross (1979-1987)
Dr. K. M. Kadish (1979-1987)
Dr. R. Kalvoda (1977-1985)
Prof. H. Kao (1981-1985)
Prof. Y. Marcus (1979-1987)
Prof. T. Mussini (1979-1987)
Prof. H. W. Nürnberg (1979-1985)
Prof. M. Senda (1979-1987)
Dr. N. Tanaka (1979-1987)
Dr. Klára Tóth (1983-1987)

National Representatives

Dr. D. D. Perrin (1971-1985)
Australia
Dr. B. Gilbert (1981-1985)
Belgium
Prof. W. C. Purdy (1979-1985)
Canada
Prof. A. A. Vlček (1984-1985)
Czechoslovakia
Prof. H. Monien (1983-1985)
Federal Republic of Germany
Dr. M. L'Her (1984-1985)
France
Dr. Gy. Farsang (1983-1985)
Hungary
Prof. H. C. Gaur (1983-1985)
India
Dr. W. F. Smyth (1981-1985)
Ireland
Prof. E. Grushka (1981-1985)
Israel
Prof. Sandra Rondinini Cavallari (1983-1985)
Italy
Dr. W. Frankvoort (1983-1985)
Netherlands
Prof. Z. Galus (1979-1985)
Poland
Prof. G. Johansson (1981-1985)
Sweden
Dr. J. Buffle (1981-1985)
Switzerland
Dr. H. Thompson (1983-1985)
United Kingdom
Prof. Janet G. Osteryoung (1981-1985)
United States of America
Prof. I. Piljac (1983-1985)
Yugoslavia

SUBCOMMITTEE ON ELECTROANALYTICAL METHODS OF ENVIRONMENTAL TRACE ANALYSIS

Prof. H. W. Nürnberg
Chairman

Members

Dr. G. E. Batley
Dr. J. Buffle
Dr. R. A. Durst
Dr. R. Kalvoda
Prof. Janet G. Osteryoung
Prof. E. Pungor
Prof. S. Rubel
Dr. W. F. Smyth

COMMISSION ON EQUILIBRIUM DATA (V.6)
(Established 1955)

Titular Members

Prof. H. Ohtaki (1977-1985)
Chairman
Dr. L. D. Pettit (1979-1987)
Secretary
Prof. E. D. Goldberg (1981-1985)
Prof. I. Grenthe (1981-1985)
Dr. I. Nagypál (1983-1987)
Dr. P. Valenta (1981-1985)

Associate Members

Prof. S. Ahrland (1983-1987)
Prof. G. Anderegg (1981-1985)
Dr. A. C. M. Bourg (1981-1985)
Dr. D. S. Gamble (1979-1987)
Dr. E. Högfeldt (1977-1985)
Prof. I. L. Khodakovskii (1983-1987)
Prof. W. A. E. McBryde (1981-1985)
Prof. G. H. Nancollas (1981-1985)
Prof. P. Paoletti (1983-1987)
Dr. D. D. Perrin (1981-1985)
Prof. D. G. Tuck (1983-1987)
Prof. O. Yamauchi (1981-1985)

National Representatives

Dr. G. T. Hefter (1983-1985)
Australia
Prof. O. E. S. Godinho (1983-1985)
Brazil
Prof. M. T. Beck (1979-1985)
Hungary
Prof. A. Bylicki (1977-1985)
Poland
Prof. C. Luca (1981-1985)
Romania
Prof. I. N. Marov (1976-1985)
Union of Soviet Socialist Republics
Prof. G. R. Choppin (1983-1985)
United States of America

SUBCOMMITTEE ON STABILITY CONSTANT DATA

Dr. L. D. Pettit
Chairman

Members

Prof. G. Anderegg
Prof. I. Grenthe
Dr. E. Högfeldt
Prof. I. L. Khodakovskii
Prof. W. A. E. McBryde
Dr. I. Nagypál
Prof. G. H. Nancollas
Prof. H. Ohtaki
Dr. D. D. Perrin
Dr. H. K. J. Powell
Prof. D. G. Tuck
Prof. O. Yamauchi

COMMISSION ON ANALYTICAL RADIOCHEMISTRY AND NUCLEAR MATERIALS (V.7)
(Under Review)

COMMISSION ON SOLUBILITY DATA (V.8)
(Established 1979)

Titular Members

Prof. A. S. Kertes (1979-1987)
Chairman
Dr. L. H. Gevantman (1983-1987)
Secretary
Prof. R. Battino (1983-1987)
Prof. H. L. Clever (1979-1987)
Prof. D. G. Shaw (1983-1987)
Dr. C. L. Young (1979-1987)

Associate Members

Prof. A. F. M. Barton (1983-1987)
Prof. R. Cohen-Adad (1979-1987)
Dr. P. G. T. Fogg (1983-1987)
Prof. F. W. Getzen (1979-1987)
Prof. W. Hayduk (1983-1987)
Prof. J. W. Lorimer (1983-1987)
Dr. A. N. Paruta (1981-1985)
Dr. M. Salomon (1983-1987)
Prof. P. W. Schindler (1979-1987)
Dr. A. Szafranski (1983-1987)
Prof. R. P. T. Tomkins (1983-1987)
Dr. S. H. Yalkowsky (1979-1987)

National Representatives

Prof. H. F. Aly (1983-1985)
Arab Republic of Egypt
Dr. R. Fernández-Prini (1980-1985)
Argentina
Dr. E. Wilhelm (1979-1985)
Austria
Dr. J. Vanderdeelen (1983-1985)
Belgium
Dr. Ch. Balarew (1980-1985)
Bulgaria
Prof. J. Fu (1983-1985)
Chinese Chemical Society
Dr. J. Hála (1984-1985)
Czechoslovakia
Dr. J. B. Jensen (1983-1985)
Denmark
Prof. B. A. Wolf (1980-1985)
Federal Republic of Germany
Dr. A. Vesala (1979-1985)
Finland
Dr. C. Treiner (1979-1985)
France
Dr. G. Jancsó (1980-1985)
Hungary
Prof. C. Kalidas (1979-1985)
India
Dr. I. Padova (1981-1985)
Israel
Prof. B. Scrosati (1980-1985)
Italy
Prof. H. Akaiwa (1980-1985)
Japan
Dr. K. H. Khoo (1983-1985)
Malaysia
Dr. G. A. Bowmaker (1980-1985)
New Zealand
Prof. H. Høiland (1981-1985)
Norway
Prof. A. Bylicki (1979-1985)
Poland
Prof. P. W. Linder (1981-1985)
Republic of South Africa
Prof. G. A. Yagodin (1979-1985)
Union of Soviet Socialist Republics
Dr. Mary Masson (1983-1985)
United Kingdom
Dr. J. D. Navratil (1981-1985)
United States of America

SUBCOMMITTEE ON GAS SOLUBILITIES

Prof. H. L. Clever
Chairman

Members

Dr. R. W. Cargill
Dr. Cynthia G. Carter
Dr. W. Gerrard
Dr. F. A. Lewis
Mr. D. M. Mason

SUBCOMMITTEE ON LIQUID SOLUBILITIES

Prof. A. F. M. Barton
Chairman

Members

Dr. Marie-Claire Haulait-Pirson
Dr. G. T. Hefter
Dr. A. L. Horvath
Prof. P. Huyskens
Dr. Zofia Maczynska
Dr. A. Maczynski
Dr. A. Skrzecz
Prof. Joan D. Willey

SUBCOMMITTEE ON SOLID SOLUBILITIES

Dr. M. Salomon
Chairman

Members

Prof. M. T. Beck
Dr. G. Bohnsack
Dr. Kathryn R. Bullock
Dr. R. W. Carling
Prof. C. Y. Chan
Dr. Marie-Thérèse Cohen-Adad
Prof. J. J. Counioux
Dr. W. Davison
Prof. T. P. Dirkse
Dr. Jitka Eysseltova
Prof. J. Fuger
Dr. C. Guminski
Dr. C. Hirayama
Dr. R. Kaliszan
Dr. I. Lambert
Dr. T. Mioduski
Prof. H. Miyamoto
Dr. R. Piekos
Dr. A. Regosz
Prof. S. Siekierski
Dr. J. A. Wood
Prof. E. M. Woolley

APPLIED CHEMISTRY DIVISION (VI)

(Established 1949)

DIVISION COMMITTEE

Dr. H. Frehse (1975-1985)
  President
Dr. P. C. Kearney (1979-1985)
  Vice-President
Dr. L. E. Coles (1981-1985)
  Secretary

Members

Dr. D. Firestone (1983-1987)
Prof. J. Holló (1981-1985)
Dr. Éva Kurucz (1981-1985)
Dipl.-Ing. K. Leichnitz (1983-1987)
Prof. M. Naudet (1981-1985)

Coopted Members

Dr. H. Egan
Dr. S. Gorbach
Dr. R. Greenhalgh
Prof. K. Kojima
Mr. S. G. Luxon

COMMISSION ON FOOD CHEMISTRY (VI.1)
(Established in its present form 1979)

Titular Members

Dr. A. E. Pohland (1983-1987)
  Chairman
Dr. P. S. Steyn (1979-1987)
  Vice-Chairman
Prof. W. Krönert (1983-1987)
  Secretary
Prof. G. Billek (1979-1985)
Prof. P. Krogh (1983-1987)
Dr. D. L. Park (1983-1987)

Associate Members

Dr. R. Battaglia (1983-1987)
Prof. M. J. J. Castegnaro (1979-1987)
Prof. G. Grimmer (1979-1987)
Prof. T. Juszkiewicz (1979-1987)
Dr. M. E. Knowles (1983-1987)
Dr. S. J. Kubacki (1983-1987)
Dr. H. A. M. G. Vaessen (1983-1987)

National Representatives

Prof. M. M. Abdel Kader (1980-1985)
  Arab Republic of Egypt
Dr. P. B. Czedik-Eysenberg (1977-1985)
  Austria
Prof. J. Davídek (1976-1985)
  Czechoslovakia
Dr. E. Lück (1975-1985)
  Federal Republic of Germany
Dr. I. Rosenthal (1983-1985)
  Israel
Dr. A. Cantafora (1977-1985)
  Italy
Dr. T. Kato (1978-1985)
  Japan
Dr. C. L. Lim (1983-1985)
  Malaysia
Prof. A. Rutkowski (1975-1985)
  Poland
Prof. C. Eriksson (1980-1985)
  Sweden
Dr. L. E. Coles (1983-1985)
  United Kingdom

COMMISSION ON BIOTECHNOLOGY (VI.2)
(Established in its present form 1979)

Titular Members

Prof. C. L. Cooney (1977-1985)
  Chairman
Prof. M. Ringpfeil (1983-1987)
  Vice-Chairman
Dr. G. G. Stewart (1983-1987)
  Secretary
Prof. T. K. Ghose (1977-1985)

Associate Members

Dr. P. P. Gray (1983-1987)
Prof. M. Linko (1983-1987)
Dr. R. C. Righelato (1983-1987)

National Representatives

Dr. R. J. Ertola (1978-1985)
  Argentina
Prof. H. J. G. Wutzel (1974-1985)
  Austria
Prof. W. Borzani (1979-1985)
  Brazil
Dr. B. Sikyta (1981-1985)
  Czechoslovakia
Prof. K. von Meyenburg (1980-1985)
  Denmark
Prof. H. Dellweg (1984-1985)
  Federal Republic of Germany
Dr. L. Penasse (1983-1985)
  France
Prof. J. Holló (1984-1985)
  Hungary
Dr. V. Jagannathan (1983-1985)
  India
Prof. L. Goldstein (1983-1985)
  Israel

Prof. F. Parisi (1982-1985)
Italy
Prof. S. Fukui (1980-1985)
Japan
Dr. B. G. Yeoh (1983-1985)
Malaysia
Prof. E. Galas (1983-1985)
Poland
Prof. Vera Johanides (1981-1985)
Yugoslavia

COMMISSION ON OILS, FATS, AND DERIVATIVES (VI.3)

(Established in its present form 1977)

Titular Members

Prof. M. Naudet (1979-1985)
Chairman
Dr. A. Hautfenne (1977-1985)
Vice-Chairman
Mr. W. D. Pocklington (1983-1987)
Secretary
Dr. Ö. Levin (1977-1985)
Dr. J. Pokorný (1977-1985)
Prof. H. Wessels (1979-1985)

Associate Members

Dr. Joyce L. Beare-Rogers (1977-1985)
Dr. A. Dieffenbacher (1981-1985)
Prof. E. Fedeli (1981-1985)
Prof. J. Gracian-Tous (1979-1985)
Mlle. Brigitte Jacobsberg (1983-1987)
Drs. F. H. Koelman (1983-1987)
Dr. Éva Kurucz (1979-1985)
Dr. R. G. Manning (1983-1987)
Mr. A. T. Møller (1979-1985)
Dr. M. Teupel (1979-1985)
Mrs. Anneli Visapää (1981-1985)
Dr. G. Zwerenz (1981-1985)

National Representatives

Prof. A. M. Gad (1980-1985)
Arab Republic of Egypt
Dr. N. G. Bazan (1980-1985)
Argentina
Dr. A. R. Johnson (1976-1985)
Australia
Dr. Regina C. A. Lago (1979-1985)
Brazil
Dr. D. G. Chobanov (1975-1985)
Bulgaria
Dr. A. P. Tulloch (1980-1985)
Canada
Dr. Gunhild Hølmer (1981-1985)
Denmark
Dr. G. Ostermann (1981-1985)
Federal Republic of Germany
Prof. R. Linko (1981-1985)
Finland
Dr. J. P. Wolff (1979-1985)
France
Prof. V. M. Kapoulas (1977-1985)
Greece
Dr. N. V. Bringi (1981-1985)
India
Dr. Mary McCarthy (1981-1985)
Ireland
Dr. A. Dolev (1983-1985)
Israel
Prof. E. Tiscornia (1982-1985)
Italy
Dr. T. Hashimoto (1974-1985)
Japan
Prof. S. H. Ong (1983-1985)
Malaysia
Mr. P. W. Hendrikse (1973-1985)
Netherlands
Dr. F. B. Shorland (1971-1985)
New Zealand
Prof. W. Zwierzykowski (1978-1985)
Poland
Mr. G. D. Carr (1980-1985)
Republic of South Africa
Dr. Rodica Stoica (1981-1985)
Romania
Prof. M. Gassiot Matas (1981-1985)
Spain
Dr. R. Ohlson (1971-1985)
Sweden
Dr. H. Brüschweiler (1981-1985)
Switzerland
Miss Reyhan Perin (1981-1985)
Turkey
Dr. V. L. Rubajlo (1981-1985)
Union of Soviet Socialist Republics
Mr. A. E. Waltking (1979-1985)
United States of America

COMMISSION ON ATMOSPHERIC ENVIRONMENT (VI.4)

(Under Review)

COMMISSION ON PESTICIDE CHEMISTRY (VI.5)

(Established in its present form 1979)

Titular Members

Mr. J. A. R. Bates (1977-1985)
Chairman
Dr. R. Greenhalgh (1975-1985)
Secretary
Dr. N. Aharonson (1983-1987)
Dr. A. Ambrus (1983-1987)
Dr. S. Gorbach (1977-1985)
Dr. W. Klein (1983-1987)

Associate Members

Dr. J. Desmoras (1977-1985)
Dr. H. O. Esser (1979-1987)
Dr. L. A. Golovleva (1983-1987)
Dr. R. J. Hemingway (1979-1987)
Dr. R. Hollingworth (1983-1987)
Dr. N. Kurihara (1983-1987)
Dr. W. B. Neely (1983-1987)
Dr. S. Otto (1983-1987)
Dr. T. R. Roberts (1979-1987)
Prof. J. Seiber (1983-1987)
Dr. D. B. Sharp (1983-1987)
Dr. J. W. Vonk (1977-1985)

National Representatives

Dr. Ana María Pechen de D'Angelo (1979-1985)
Argentina
Dr. W. Lara (1983-1985)
Brazil
Dr. Z. Li (1983-1985)
Chinese Chemical Society
Dr. J. Kovacicova (1983-1985)
Czechoslovakia
Dr. F. Dutka (1979-1985)
Hungary
Dr. S. K. Mukerjee (1983-1985)
India
Dr. P. Bracha (1983-1985)
Israel
Dr. J. Miyamoto (1984-1985)
Japan
Dr. C. K. Heng (1983-1985)
Malaysia
Dr. G. W. Mason (1975-1985)
New Zealand
Dr. A. Kotarski (1979-1985)
Poland
Prof. N. Bărbulescu (1981-1985)
Romania
Dr. P. C. Kearney (1983-1985)
United States of America

COMMISSION ON WATER QUALITY (VI.6)
(Under Review)

CLINICAL CHEMISTRY DIVISION (VII)

(Established in its present form 1979)

DIVISION COMMITTEE

Prof. F. W. Sunderman, Jr. (1979-1985)
President
Dr. S. S. Brown (1979-1985)
Vice-President
Prof. N. Montalbetti (1981-1985)
Secretary

Members

Dr. A. Kallner (1981-1985)
Prof. Mirna Mikač-Dević (1981-1985)
Prof. K. Okuda (1981-1985)
Dr. V. N. Titov (1981-1985)
Prof. D. J. Vonderschmitt (1981-1985)

Coopted Members

Prof. G. Férard
Dr. C. Franzini
Prof. G. Hoxter
Dr. K. S. Kasprzak
Dr. S. Nomoto
Prof. B. Ribas
Mr. J. C. Rigg
Prof. R. Vihko

National Representatives

Dr. J. C. Connelly (1983-1985)
Australia
Prof. E. Kaiser (1969-1985)
Austria
Dr. C. K. Harris (1979-1985)
Canada
Dr. U. Worsaae (1975-1985)
Denmark
Prof. J. Breuer (1983-1985)
Federal Republic of Germany
Prof. J. C. Jardillier (1979-1985)
France
Dr. Maroulio J. Talieri-Gianopoulou (1983-1985)
Greece
Dr. B. Ringelhann (1973-1985)
Hungary
Prof. M. Cais (1981-1985)
Israel
Prof. G. D'Ascenzo (1983-1985)
Italy
Prof. Y. Yamamura (1967-1985)
Japan
Dr. S. Landaas (1983-1985)
Norway
Prof. S. Angielski (1978-1985)
Poland
Prof. S. Lindstedt (1973-1985)
Sweden
Prof. V. N. Orekhovich (1967-1985)
Union of Soviet Socialist Republics
Prof. P. D. Griffiths (1979-1985)
United Kingdom

Representatives of IFCC

Dr. R. Dybkaer
President
Dr. J. G. Hill
Secretary

Representative of IUB

Prof. P. N. Campbell

COMMISSION ON AUTOMATION AND CLINICAL CHEMICAL TECHNIQUES (VII.1)
(Established in its present form 1981)

Titular Members

Prof. R. Haeckel (1981-1985)
Chairman
Dr. T. D. Geary (1983-1987)
Secretary
Prof. P. Bonini (1983-1987)
Dr. C. A. Burtis (1981-1985)

Associate Members

Dr. J. Bierens de Haan (1983-1987)
Prof. J. Büttner (1981-1985)
Prof. M. Hjelm (1983-1987)
Dr. D. S. Young (1981-1985)

National Representatives

Dr. R. D. Baillie (1981-1985)
Canada
Prof. P. Valdiguié (1981-1985)
France
Dr. V. Haviaras (1983-1985)
Greece
Prof. F. Salvatore (1981-1985)
Italy
Dr. Ch. Hayashi (1981-1985)
Japan
Mrs. Sissel Strand (1981-1985)
Norway
Dr. A. Hamfelt (1981-1985)
Sweden
Dr. N. Gochman (1981-1985)
United States of America

COMMISSION ON QUANTITIES AND UNITS (VII.2)
(Established 1967)

Titular Members

Dr. H. P. Lehmann (1979-1987)
Chairman
Dr. D. R. Bangham (1983-1987)
Prof. L. F. Bertello (1983-1987)
Dr. G. Férard (1983-1987)
Dr. B. F. Visser (1977-1985)

Associate Members

Dr. R. Dybkaer (1977-1985)
Dr. C. Onkelinx (1983-1987)
Prof. O. Siggaard-Andersen (1983-1987)
Dr. U. Worsaae (1983-1987)
Dr. R. Zender (1981-1987)

National Representatives

Dr. D. J. Campbell (1981-1985)
Canada
Prof. D. Stamm (1981-1985)
Federal Republic of Germany
Prof. B. Bousquet (1982-1985)
France
Dr. G. Sotiropoulou (1983-1985)
Greece
Prof. N. Montalbetti (1982-1985)
Italy
Prof. T. Horio (1981-1985)
Japan
Dr. O. P. Foss (1982-1985)
Norway
Prof. C.-H. de Verdier (1981-1985)
Sweden
Prof. P. D. Griffiths (1981-1985)
United Kingdom
Dr. C. A. Burtis (1981-1985)
United States of America

COMMISSION ON TEACHING (VII.3)
(Established 1967)

Titular Members

Dr. M. K. Schwartz (1975-1985)
Chairman
Dr. H. G. J. Worth (1979-1987)
Secretary
Dr. Nelly de Cediel (1983-1987)
Dr. C. G. Fraser (1981-1985)
Dr. O. Zinder (1979-1987)

Associate Members

Dr. Jocelyn Hicks (1983-1987)
Prof. N. Montalbetti (1983-1987)
Dr. C. J. Porter (1983-1987)
Dr. M. Roth (1981-1985)
Prof. B. Štraus (1979-1987)
Dr. V. N. Titov (1981-1985)
Prof. R. Vihko (1981-1985)
Dr. C. W. H. Walker (1983-1987)

National Representatives

Prof. M. M. Abdel Kader (1980-1985)
Arab Republic of Egypt
Dr. A. Gornall (1981-1985)
Canada
Prof. A. F. Delbrück (1981-1985)
Federal Republic of Germany
Prof. J. Agneray (1982-1985)
France
Dr. A. G. Hadjivassiliou (1983-1985)
Greece
Prof. R. Strom (1982-1985)
Italy
Prof. T. Kanno (1981-1985)
Japan
Dr. B. Christophersen (1981-1985)
Norway
Dr. B. Kågedal (1981-1985)
Sweden
Dr. H. A. Fritsche, Jr. (1981-1985)
United States of America

COMMISSION ON TOXICOLOGY (VII.4)
(Established 1973)

Titular Members

Prof. J. Savory (1977-1985)
Chairman
Prof. D. B. Tonks (1977-1985)
Secretary
Prof. Mirna Mikač-Devič (1981-1985)
Dr. E. Nieboer (1983-1987)
Dr. M. Stoeppler (1979-1985)
Dr. Y. Thomassen (1983-1987)

Associate Members

Prof. A. Aitio (1983-1987)
Dr. K. S. Kasprzak (1981-1985)
Dr. K. Kühn (1981-1985)
Dr. S. Nomoto (1981-1985)
Prof. W. C. Purdy (1981-1985)
Dr. D. Shapcott (1983-1987)

National Representatives

Dr. B. M. Kapur (1981-1985)
Canada
Prof. J. Büttner (1981-1985)
Federal Republic of Germany
Prof. C. Bohuon (1981-1985)
France
Mr. C. Kotsovassilis (1983-1985)
Greece
Dr. J. Borsy (1980-1985)
Hungary
Dr. W. King (1983-1985)
Ireland
Prof. F. Lodi (1983-1985)
Italy
Prof. T. Nambara (1980-1985)
Japan
Dr. W. A. Temple (1983-1985)
New Zealand

Dr. L. Daae (1980-1985)
Norway
Prof. B. Sörbo (1980-1985)
Sweden
Dr. P. I. Jatlow (1981-1985)
United States of America

Representative of IUPHAR Section on Toxicology

Prof. G. Plaa

SUBCOMMITTEE ON ENVIRONMENTAL AND OCCUPATIONAL TOXICOLOGY OF NICKEL

Dr. E. Nieboer
Chairman

Members

Prof. A. Aitio
Mr. I. Andersen
Prof. M. Anke
Dr. I. Balogh
Dr. M. Boysen
Mr. Ch. Bozec
Dr. S. S. Brown
Dr. D. D. Bryson
Dr. A. G. Cecutti
Dr. J. M. Christensen
Dr. M. Costa
Dr. R. De Neve
Dr. D. Dewally
Dr. A. Dornemann
Dr. V. Eybl
Dr. R. Germain
Dr. Ph. Grandjean
Dr. K. Hansen
Dr. H. F. Hildebrand
Dr. A. Høgetveit
Dr. J. Järvisalo
Dr. K. S. Kasprzak
Dr. K. Kühn
Dr. Joan C. McEwan
Dr. E. Mastromatteo
Dr. J. Meininger
Prof. Mirna Mikač-Devič
Dr. L. G. Morgan
Dr. P. Mushak
Dr. S. Nomoto
Dr. T. Norseth
Dr. J. Olafsson
Dr. A. K. Reith
Mr. B. R. Roy
Dr. G. Rubanyi
Dr. E. Sabbioni
Dr. B. Sarkar
Mrs. Geraldine Savory
Prof. J. Savory
Dr. K. H. Schaller
Dr. J. A. A. Sjöholm
Dr. R. M. Stern
Dr. M. Stoeppler
Prof. F. W. Sunderman, Jr.
Prof. F. W. Sunderman, Sr.
Dr. S. K. Tandon
Dr. H. Tjalve
Dr. Y. Thomassen
Prof. D. B. Tonks
Prof. D. R. Williams
Dr. H. Zachariasen

SUBCOMMITTEE ON ENVIRONMENTAL AND OCCUPATIONAL TOXICOLOGY OF CADMIUM

Dr. M. Stoeppler
Chairman
Dr. H. T. Delves
Secretary

Members

Dr. J. Angerer
Dr. J. Aretz
Mr. M. Bédard
Mrs. E. Berman
Dr. S. S. Brown
Dr. A. Buratti
Mrs. J. M. Christensen
Prof. P. De Bièvre
Dr. R. De Neve
Dr. A. Dornemann
Dr. V. Ebbighausen
Dr. U. Ewers
Dr. G. S. Fell
Dr. M. Fleischer
Dr. J. P. Franke
Prof. J. M. Frazier
Prof. S. Gomiščeк
Dr. R. M. Griffin
Mrs. C. Hadtzchristidis
Prof. J. M. Haguenoer
Dr. A. Harada
Dr. R. F. M. Herber
Dr. F. Hill
Mrs. V. Hudnik
Mr. Jakobi
Dr. J. Järvisalo
Dr. S. Khandelwal
Prof. Y. Kodama
Dr. P. Krause
Prof. R. Lauwerys
Mr. B. Lind
Prof. P. Lous
Prof. G. Machata
Dr. G. Mattiello
Mr. H. Muntau
Dr. H. Nomiyama
Dr. S. Nomoto
Dr. H. Oldiges
Prof. J. M. Ottaway
Prof. H.-J. Pesch
Dr. J. M. Pujade-Renaud
Dr. D. A. Robbins
Prof. S. Rubel
Dr. K. H. Schaller
Mr. N. J. van Sittert
Mr. N. J. Smith
Dr. K. R. Sperling

Prof. W. Steuer
Dr. C. D. Strehlow
Dr. K. S. Subramanian
Prof. D. B. Tonks
Prof. J. Versieck
Prof. O. Vesterberg
Mrs. R. J. Ward
Dr. E. Weisenberg
Dr. W. C. M. Zwennis

SUBCOMMITTEE ON TOXICOLOGY OF ALUMINIUM

Prof. J. Savory
Chairman

Members

Dr. A. C. Alfrey
Prof. P. Allain
Prof. A. Berlin
Dr. U. de Boni
Dr. E. Claiborne Irby
Prof. H. A. Das
Dr. J. P. Day
Dr. G. S. Fell
Dr. H. J. Gitelman
Dr. F. Y. Leung
Dr. D. R. C. McLachlan
Dr. H. Matusiewicz
Dr. Y. Mauras
Dr. E. Nieboer
Dr. B. Stevens
Dr. A. Taylor
Dr. M. K. Ward
Dr. M. R. Wills

SUBCOMMITTEE ON TOXICOLOGY OF SELENIUM

Dr. Y. Thomassen
Chaiman

Members

Dr. J. Aaseth
Dr. G. Alfthan
Dr. D. Behne
Prof. H. A. Das
Dr. H. T. Delves
Dr. R. M. Griffin
Dr. M. Ihnat
Dr. V. Iyengar
Dr. O. A. Levander
Prof. Mirna Mikač-Devic
Dr. J. Pikkarainen
Dr. Marion F. Robinson
Dr. M. Stoeppler
Dr. Christine D. Thomson
Dr. R. T. Tulley
Dr. C. Veillon
Dr. Marleen Verlinden
Dr. J. H. Watkinson
Dr. B. Welz
Dr. G. Q. Yang

STANDING ORDERS FROM EXECUTIVE COMMITTEE

* CHEMRAWN COMMITTEE

Composition and Terms of Office

(i) There shall be a standing CHEMRAWN Committee, composed of a Chairman, a Secretary, six regular Members, and up to six Associate Members.

(ii) The President in consultation with the Executive Committee shall appoint the Members and the Associate Members. The CHEMRAWN Committee may propose names of persons suitably qualified for appointment.

(iii) The period of service of a regular Member and of an Associate Member shall be four years, renewable to a maximum of eight years.

(iv) The Membership shall be reviewed every two years by the incoming President in consultation with the Executive Committee.

(v) The President in consultation with the Executive Committee shall appoint the Chairman and the Secretary. The CHEMRAWN Committee may propose candidates.

(vi) The period of service of the Chairman and of the Secretary shall not exceed eight years. The sum of the years of service as a regular Member or an Associate Member and as the Chairman or the Secretary shall not exceed ten years.

Terms of Reference

On behalf of the President and the Executive Committee:

(i) to identify human needs amenable to solution through chemistry with particular attention to those areas of global or multinational interest,

(ii) to serve as an international body and forum for the gathering, discussion, advancement, and dissemination of chemical knowledge deemed useful for the improvement of man and his environment,

(iii) to serve as an international, nongovernmental source of advice for the benefit of governments and international agencies with respect to chemistry and its application to world needs,

and to be responsible for organizing IUPAC activity in these areas as approved by the President and the Executive Committee.

* CHEMRAWN = CHEMical Research Applied to World Needs

COMMITTEE ON CHEMISTRY AND INDUSTRY (COCI)

Composition and Terms of Office

(i) There shall be a standing Committee on Chemistry and Industry, composed of a Chairman, a Secretary, and a Member from each country or region having more than one Company Associate.

(ii) The President in consultation with the Executive Committee shall appoint the Chairman, the Secretary, and the Members. Candidates may be proposed by the Committee on Chemistry and Industry and by National Adhering Organizations having Company Associate programmes.

(iii) The period of service shall be four years, renewable for a further term of four years.

Terms of Reference

On behalf of the President and the Executive Committee:

(i) to guide and inform Company Associates on IUPAC programmes and policies,

(ii) to establish important areas of chemical science in IUPAC related to ongoing industrial needs,

(iii) to put to industry enquiries on major chemical matters.

COMMITTEE ON PUBLICATIONS

Composition and Terms of Office

(i) There shall be a standing Committee on Publications, composed of a Chairman, at least three and not more than six other Members, and the Scientific Editor of Pure and Applied Chemistry (PAC).

(ii) The President in consultation with the Executive Committee shall appoint the Members. The Committee on Publications may propose names of persons suitably qualified for appointment.

(iii) The maximum period of service of the ordinary Members shall be eight years.

(iv) The Membership shall be reviewed every two years by the incoming President in consultation with the Executive Committee.

(v) The President in consultation with the Executive Committee shall appoint the Chairman. The Committee on Publications may propose candidates.

(vi) The period of service of the Chairman shall not exceed eight years. The sum of the years of service as an ordinary Member and as the Chairman shall not exceed ten years.

Terms of Reference

(i) To advise the President and the Executive Committee on all matters of publications including policy.

(ii) To make recommendations for decision by the President and/or the Executive Committee.

CHEMISTRY INTERNATIONAL EDITORIAL ADVISORY BOARD

Composition and Terms of Office

(i) There shall be an Editorial Advisory Board for the IUPAC news magazine Chemistry International (CI).

(ii) The Board shall consist of the Secretary General (Chairman), the Chairman of the Committee on Publications, one representative from each of the Divisions and from the CHEMRAWN Committee, the Committee on Chemistry and Industry, the Committee on Teaching of Chemistry, the Interdivisional Committee on Nomenclature and Symbols, and the CI Editor (Secretary).

(iii) The representatives shall be nominated biennially by the President or the Chairman of the relevant IUPAC bodies and be appointed by the President of the Union in consultation with the Executive Committee.

Terms of Reference

(i) To review the editorial policy and contents of CI at a General Assembly.

(ii) To advise the Editor, upon request, on articles to appear in CI.

(iii) To discuss future contents and format.

(iv) To ensure that IUPAC bodies provide the Editor, spontaneously or on request, with material on their work and future programmes.

COMMITTEE ON TEACHING OF CHEMISTRY (CTC)

Composition and Terms of Office

(i) There shall be a standing Committee on Teaching of Chemistry, composed of a Chairman, a Secretary, and six other Members.

(ii) The President in consultation with the Executive Committee shall appoint the Members. The Committee on Teaching of Chemistry may propose names of persons suitably qualified for appointment.

(iii) The maximum period of service of the ordinary Members shall be eight years.

(iv) The Membership shall be reviewed every two years by the incoming President in consultation with the Executive Committee.

(v) The President in consultation with the Executive Committee shall appoint the Chairman and the Secretary. The Committee on Teaching of Chemistry may propose candidates.

(vi) The period of service of the Chairman and of the Secretary shall not exceed eight years. The sum of the years of service as an ordinary Member and as the Chairman or the Secretary shall not exceed ten years.

Terms of Reference

(i) To advise the President and the Executive Committee on educational matters.

(ii) To coordinate the educational interests and activities of all IUPAC bodies.

(iii) To act as an informational and coordinating body for chemical education activities throughout the world and to establish a system of National Representatives as a channel of communication. To develop liaison with international organizations such as UNESCO and with national chemical societies or chemical education committees.

FINANCE COMMITTEE (FC)

Composition and Terms of Office

(i) There shall be a standing Finance Committee, composed of five Members and a Chairman. In addition, the Treasurer shall be an ex officio Member but without voting power.

(ii) The President in consultation with the Executive Committee shall appoint the Members. The Finance Committee may propose names of persons suitably qualified for appointment.

(iii) The maximum period of service of the ordinary Members shall be eight years.

(iv) The Membership shall be reviewed every two years by the incoming President in consultation with the Executive Committee.

(v) The President in consultation with the Executive Committee shall appoint the Chairman. The Finance Committee may propose candidates.

(vi) The period of service of the Chairman shall not exceed eight years. The sum of the years of service as an ordinary Member and as the Chairman shall not exceed ten years.

Terms of Reference

(i) To advise the President and the Executive Committee on financial matters.

(ii) To make recommendations for decision by the President and/or the Executive Committee.

(iii) To review the IUPAC investment portfolio at least annually and to make such changes as appear appropriate.

(iv) The Finance Committee shall not have executive functions except with respect to dealings in securities. The Finance Committee shall have executive authority with respect to selection, purchases, and sales of securities held by IUPAC, provided that the Treasurer and the IUPAC Banker concur with the decisions of the Finance Committee.

INTERDIVISIONAL COMMITTEE ON NOMENCLATURE AND SYMBOLS (IDCNS)

Composition and Terms of Office

(i) There shall be a standing Interdivisional Committee on Nomenclature and Symbols, composed of a Chairman, a Secretary, the Scientific Editor of PAC, one Member and one Associate Member from each Division.

(ii) The President in consultation with the Executive Committee shall appoint the Members and the Associate Members of the Divisions. IDCNS and the Division Presidents may propose names of persons suitably qualified for appointment.

(iii) The period of service of the Members and of the Associate Members shall be four years, renewable for a further term of four years.

(iv) The President in consultation with the Executive Committee shall appoint the Chairman and the Secretary. IDCNS may propose candidates. The President in consultation with the Chairman may designate up to three additional experts at IUPAC expense for specific business at a meeting of IDCNS.

(v) The period of service of the Chairman and of the Secretary shall be four years, renewable for a further term of four years. The sum of the years as a Member or as an Associate Member and as the Chairman or the Secretary shall not exceed ten years.

(vi) The following organizations shall be invited to attend meetings of IDCNS:

Bureau International des Poids et Mesures
International Organization for Standardization
International Union of Biochemistry
International Union of Nutritional Science
International Union of Pure and Applied Biophysics
International Union of Pure and Applied Physics

Terms of Reference

(i) To be responsible for submission to the Bureau/Council, in accordance with Bylaw 2.11, for publication or otherwise, any IUPAC document concerned with nomenclature and symbols.

(ii) Before recommending any material for publication as an IUPAC document, to ensure that the fullest possible consultations have taken place, and the widest possible consensus has been reached among all Divisions, Commissions, and other bodies of the Union, and between IUPAC and other ICSU bodies, the international standardizing organizations, and Conférence Générale des Poids et Mesures (CGPM) and its Committees.

(iii) To ensure that any considered IUPAC view shall carry the fullest possible weight among other international organizations, all negotiations on matters concerned with nomenclature and symbols with other ICSU bodies, with the international standardizing organizations, and with CGPM and its Committees, shall be conducted through IDCNS, which shall advise the Executive Committee accordingly.

(iv) To be responsible, after consultation with all relevant bodies of IUPAC, for the official IUPAC comments on all documents on nomenclature and symbols sent to the Union for comment.

(v) To advise the President and the Executive Committee on suitable persons for appointment as representatives of IUPAC on other bodies concerned with nomenclature and symbols.

IUB-IUPAC JOINT COMMISSION ON BIOCHEMICAL NOMENCLATURE (JCBN)

Composition and Terms of Office

(i) There shall be a standing IUB-IUPAC Joint Commission on Biochemical Nomenclature, composed of eight regular Members, one of whom shall serve as the Chairman and one as the Secretary, and up to eight Associate Members.

(ii) The Chairman and the Secretary shall each be appointed for a term of three years by the Presidents of the two Unions, subject to whatever ratification is imposed by Union Statutes and Bylaws. JCBN may propose names of persons suitably qualified for appointment.

(iii) Four of the regular Members shall be appointed by IUB and four by IUPAC. One of the four IUB regular Members shall be an Officer of the IUB Committee of Editors of Biochemical Journals (CEBJ). Four of the Associate Members shall be appointed by IUB and four by IUPAC.

(iv) The regular Members appointed by IUB shall also be Members of the Nomenclature Committee of IUB (NC-IUB).

(v) Candidates for membership may be proposed by the Commission and shall be appointed by the respective Unions.

(vi) The periods of service of the regular Members and of the Associate Members shall be in accordance with the Statutes and Bylaws of the appointing Union. The sum of the years of service as a regular Member and as the Chairman or the Secretary shall not exceed ten years.

Terms of Reference

(i) To consider the nomenclature and symbolic representation of natural products and related compounds including synthetic analogues, and, where appropriate, to make recommendations in conformity with those issued by other nomenclature Commissions of IUB and IUPAC.

(ii) Before recommending any material for publication as an IUB-IUPAC document, to ensure that the fullest possible consultations have taken place and the widest possible consensus has been reached with the appropriate bodies of the two Unions. For IUPAC, these are the Interdivisional Committee on Nomenclature and Symbols (IDCNS), or, through IDCNS, the appropriate divisional nomenclature Commissions. For IUB, these are NC-IUB and CEBJ.

(iii) Approval to publish any material as an IUB-IUPAC document is to be obtained, in the case of IUPAC from IDCNS, and in the case of IUB from the Executive Committee of IUB.

(iv) The Commission shall normally hold an annual meeting, concurrently with NC-IUB, and with the approval of the President of IUPAC and the Executive Committee of IUB.

(v) Associate Members will receive all documents of the Commission and their opinion will be sought by correspondence. An Associate Member may attend any meeting of the Commission, but his expenses will not be defrayed by the respective Executive Committee unless he has been invited as an observer under item (vi).

(vi) The Commission shall be entitled to invite observers from other Commissions and experts in special fields to attend the meeting. The respective Executive Committee shall be asked in advance to defray any expenses that would arise from such an invitation.

(vii) The continuation of JCBN will be reviewed by the Executive Committees of the two Unions in 1986.

IUB-IUPAB-IUPAC COMMISSION ON BIOTHERMODYNAMICS (CB)

Composition and Terms of Office

(i) There shall be a standing IUB-IUPAB-IUPAC Commission on Biothermodynamics, composed of a Chairman, five other regular Members, and up to six Associate Members.

(ii) Each Union, in consultation with the Commission, shall appoint two of the regular Members. The Commission shall appoint its Associate Members. The regular Members shall elect the Chairman from among themselves.

(iii) The initial period of appointment of the regular Members shall be three years. Appointments shall be renewable for additional three-year terms up to a maximum of nine years.

(iv) The period of appointment of the Associate Members shall be two years. Appointments of the Associate Members shall be renewable for additional terms with no set limits.

Terms of Reference

(i) To advise the Unions on matters relating to the applications of thermodynamics in biology; in particular, to survey the status of research in this field, to promote the compilation and evaluation of relevant thermodynamic data and their correct presentation by thermodynamic functions, and to appraise the value of thermodynamic information for the development of biology.

(ii) To make recommendations on nomenclature in biothermodynamics and to formulate guidelines for the presentation of thermodynamic data relating to biological systems.

(iii) To initiate, sponsor, and organize meetings and conferences on biothermodynamic topics, both within the existing framework of the Union meetings and as independent activities. To act as a central source of information on meetings, workshops, and training courses dealing with research and education in biothermodynamics, and to act as a means of coordinating and arranging such activities.

(iv) To encourage and participate actively in the preparation and publication of monographs in the field of biothermodynamics.

ALPHABETICAL INDEX TO MEMBERS OF IUPAC BODIES 1983 — 1985

Where telephone numbers are quoted, the country code, area code in brackets, and subscriber's number are prefixed by a plus sign and this should be substituted by a member's own international country code, which varies from country to country. Telex numbers with answerback codes are included where these have been supplied by members. The numbers in brackets at the end of each entry indicate the pages on which the memberships of IUPAC bodies for each individual will be found.

So that the female members of the Union can be identified, their forenames have been printed in capital letters.

> If members who have not already provided the Secretariat with their telephone and telex numbers would do so, the records in Oxford could be completed. Also, will members please advise the Secretariat of any errors; and changes which are necessary.

AASETH, Dr. J., Institute of Occupational Health, Gydas Vei 8, POB 8149 Dep., Oslo 1, Norway
(TEL: +47 (2) 466850) (38)

ABDEL KADER, Prof. M. M., Nutrition Institute, 16 Kasr El-Ainy Street, Cairo, Arab Republic of Egypt (32, 36)

ABEL, Prof. E., Department of Chemistry, University of Exeter, Stocker Road, Exeter EX4 4QD, UK
(TEL: +44 (392) 77911) (6)

ACEVEDO, Prof. R., c/o Facultad de Quimica, Instituto Superior Pedagogico "Enrique Jose Varona", Ciudad Libertad, Marianao, Habana 5, Cuba (11)

ACHMATOWICZ, Prof. O., Instytut Chemii i Chemii Rolniczej, Akademia Rolnicza w Warszawie, ul. Rakowiecka 26-30, PL-02 528 Warszawa, Poland
(TEL: +48 (22) 492251. TELEX: 814790 PL) (21)

ACKERMANN, Prof. G., Lehrstuhl für Analytische Chemie, Sektion Chemie, Bergakademie Freiberg, Leipziger Strasse, DDR-9200 Freiberg, German Democratic Republic
(TEL: +37 (51) 2325. TELEX: 785035 BAF DD) (27)

ADDISON, Prof. C. C., Department of Chemistry, University of Nottingham, University Park, Nottingham NG7 2RD, UK
(TEL: +44 (602) 56101) (12)

AGAR, Dr. J. N., Department of Physical Chemistry, University of Cambridge, Lensfield Road, Cambridge CB2 1EP, UK
(TEL: +44 (223) 66499. TELEX: 81240 CAMSPL G) (16)

AGIUS, Dr. P. J., Thickets, Hinksey Hill Top, Oxford OX1 5BQ, UK
(TEL: +44 (865) 735119. TELEX: 83147 G VIA OR AGIUS) (12)

AGNERAY, Prof. J., Centre d'Etudes Pharmaceutiques et Biologique, Laboratoire de Biochimie, Université Paris Sud, Tour D 4, 1er étage, rue Jean Baptiste Clément, F-92290 Chatenay Malabry, France (36)

AGRANAT, Prof. I., Institute of Chemistry, Hebrew University of Jerusalem, Jerusalem 91904, Israel
(TEL: +972 (2) 35381. TELEX: 25391 HU IL) (22)

AHARONSON, Dr. N., Institute of Plant Protection, Agricultural Research Organization, Ministry of Agriculture, Volcani Center, POB 6, Bet Dagan 50250, Israel
(TEL: +972 (3) 940440. TELEX: 341118 BXTV IL) (33)

AHLBERG, Prof. P. N. I., Institute of Chemistry, Department of Organic Chemistry, University of Uppsala, Box 531, S-751 21 Uppsala, Sweden
(TEL: +46 (18) 139460. TELEX: 76076 S) (22)

AHLUWALIA, Prof. J. C., Department of Chemistry, Indian Institute of Technology, New Delhi-110016, India (16)

AHRLAND, Prof. S., Department of Inorganic Chemistry 1, Chemical Center, University of Lund, Box 740, S-220 07 Lund 7, Sweden
(TEL: +46 (46) 108105. TELEX: S-33533 LUNIVER) (29)

AITIO, Prof. A., Institute of Occupational Health, Department of Industrial Hygiene and Toxicology, Haartmaninkatu 1, SF-00290 Helsinki 29, Finland
(TEL: +358 (0) 47471. TELEX: 121394 OCCUPHEALTH HELSINKI SF) (36, 37)

AJROLDI, Dr. G., Montepolimeri SpA, Centre Ricerche Bollate, Via S. Pietro 50, I-20021 Bollate, Italy
(TEL: +39 (2) 350 1201. TELEX: 310679 MONTED I) (24)

AKAIWA, Prof. H., Department of Technology, Faculty of Engineering, Gunma University, 1-5-1 Tenjin-cho, Kiryu, Gunma, 376 Japan
(TEL: +81 (277) 22 3181) (30)

ALBERTY, Prof. R. A., Department of Chemistry, Massachusetts Institute of Technology, Cambridge, Massachusetts 02139, USA
(TEL: +1 (617) 253 2456. TELEX: 92 1473 MITCAM) (15)

ALCOCK, Prof. C. B., Department of Metallurgy and Materials Science, University of Toronto, Toronto, Ontario M5S 1A4, Canada
(TEL: +1 (416) 978 6808. TELEX: 06-23887 CA) (20)

ALFREY, Dr. A. C., Renal Division, Denver Veterans Administration Hospital, 1055 Clermont Street, Denver, Colorado 80220, USA
(TEL: +1 (303) 399 8020) (38)

ALFTHAN, Dr. G., National Public Health Institute, Mannerheimintie 166, SF-00280 Helsinki 28, Finland (38)

ALKEMADE, Prof. C. Th. J., Fysisch Laboratorium, Rijksuniversiteit Utrecht, Princetonplein 5, 3584 CC Utrecht, Netherlands
(TEL: +31 (30) 533302) (28)

ALLAIN, Prof. P., Laboratoire de Pharmacologie, C.H.U., F-49036 Angiers, France
(TEL: +33 (41) 88 69 51) (38)

ALLEGRA, Prof. G., Istituto di Chimica, Politecnico di Milano, Piazza Leonardo da Vinci 32, I-20133 Milano, Italy
(TEL: +39 (2) 230845/9. TELEX: 333467 POLIMI I) (23)

ALLEN, Prof. Sir Geoffrey, Unilever PLC, Unilever House, POB 68, Blackfriars, London EC4P 4BQ, UK
(TEL: +44 (1) 822 6399) (24)

ALLSOPP, Dr., Plastics Division, Imperial Chemical Industries PLC, POB 6, Bessemer Road, Welwyn Garden City, Hertfordshire AL7 1HD, UK
(TEL: +44 (7073) 23400. TELEX: 264251 ICIPLAST WELWYN G) (25)

ALY, Prof. H. F., 20 El-Nour Street, Dokki, Giza 12611, Arab Republic of Egypt (30)

AL-NIAIMI, Prof. N. S., Federation of Asian Chemical Societies, c/o Iraqi Chemical Society, POB 8011, Salihiyah, Baghdad, Iraq (5)

AMBRUS, Dr. A., Mezogazdasagi és Elelmezesügyi Miniszterium, Növenyvedelmi és Agrokemiai Kozpont, XI Budaorsi utca 141-145, Pf 127, H-1502 Budapest, Hungary
(TEL: +36 (1) 259 540. TELEX: 226906 H) (33)

AMES, Dr. S. R., 61 Biltmore Drive, Rochester, New York 14617, USA (12)

AMOUROUX, Prof. J., Laboratoire de Génie Chimique, Ecole Nationale Supérieure de Chimie, Université Pierre et Marie Curie, 11 rue Pierre et Marie Curie, F-75231 Paris Cédex 05, France
(TEL: +33 (1) 336 25 25/329 12 21) (15)

ANDEREGG, Prof. G., Laboratorium für Anorganische Chemie der Eidgenössischen Technischen Hochschule Zürich, Universitätstrasse 6, CH-8092 Zürich, Switzerland
(TEL: +41 (1) 256 28 53. TELEX: 53178 ETHBI CH) (29, 30)

ANDERSEN, Mr. I., Falconbridge Nikkelverk Aktieselskap, POB 457, N-4601 Kristiansand S, Norway
(TEL: +47 (42) 25500) (37)

ANDERSON, Dr. B. C., Central Research and Development Department, E. I. du Pont de Nemours & Co. Inc., Experimental Station, Building 328, Wilmington, Delaware 19898, USA
(TELEX: 835420 DUPONT) (23, 24)

ANDERSON, Dr. H. R., Department 49F, IBM, Building 300-40E, East Fishkill Facility, Route 52, Hopewell Junction, New York 12533, USA (26)

ANGERER, Dr. J., Zentralinstitut für Arbeitsmedizin, Adolph-Schönfelder-Strasse 5/XI, D-2000 Hamburg 75, Federal Republic of Germany
(TEL: +49 (40) 291882774) (37)

ANGIELSKI, Prof. S., Zakład Biochemii Klinicznej, Instytut Patologii Akademii Medycznej, ul. Dębniki 7, PL-80 211 Gdańsk, Poland (35)

ANGUS, Dr. S., Department of Chemical Engineering and Chemical Technology, Imperial College of Science and Technology, Prince Consort Road, South Kensington, London SW7 2BY, UK
(TEL: +44 (1) 589 5111. TELEX: 261503 G) (15, 16)

ANKE, Prof. M., Wissenschaftsbereich Tierernährungschemie der Karl-Marx-Universität, Dornburger Strasse 24, DDR-6900 Jena, German Democratic Republic
(TEL: +37 (791) 411112) (37)

ANSARA, Dr. I., Laboratoire de Thermodynamique et Physico-Chimie Métallurgiques associé au Centre National de la Recherche Scientifique, Domaine Universitaire, ENSEEG-INPG, BP 44, F-38401 Saint Martin d'Hères, France
(TEL: +33 (76) 54 41 27. TELEX: 980145 THERMO SMHER F) (15)

ANTHONY, Dr. ANNE-MARIE, Centre de Recherches sur la Physique des Hautes Températures du Centre National de la Recherche Scientifique, 1D avenue de la Recherche Scientifique, F-45045 Orléans Cédex, France
(TEL: +33 (38) 63 06 17. TELEX: 760351 CNRSORL F) (20)

ARETZ, Dr. J., Abteilung Hygiene und Arbeitsmedizin der Medizinischen Fakultät, RWTH Aachen, Lochnerstrasse 4-20, D-5100 Aachen, Federal Republic of Germany
(TEL: +49 (241) 8088880) (37)

ARMSTRONG, Prof. R. C., Department of Chemical Engineering, Room 66-505, Massachusetts Institute of Technology, Cambridge, Massachusetts 02139, USA
(TEL: +1 (617) 253 4581. TELEX: 92 1473 MITCAM) (24, 25)

ARVÍA, Dr. A. J., Instituto de Investigaciones Fisicoquímicas, Teóricas y Aplicadas (INIFTA), Casilla de Correo 16, Sucursal 4, 1900 La Plata, Argentina
(TEL: +54 (21) 33204) (10, 16)

AVNI, Prof. R., Nuclear Research Center Negev, POB 9001, Beer-Sheva 84160, Israel
(TEL: +972 (57) 58080) (15)

BADRI, Dr. M., Department of Chemistry, Universiti Pertanian Malaysia, Serdang, Selangor, Malaysia
(TEL: +60 (3) 356101. TELEX: 37454 UNIPERTA MA) (20)

BAILLIE, Dr. R. D., Department of Laboratory Medicine, Victoria General Hospital - Helmcken, 35 Helmcken Road, Victoria, British Columbia V8Z 6R5, Canada
(TEL: +1 (604) 727 4167. TELEX: 0497267 CA) (35)

BALABAN, Prof. A. T., Strada Poenaru Bordea 6, apart. 13, R-70502 Bucureşti, Romania
(TEL: 146455. TELEX: 10490 PI BUH R) (21, 22)

BALAREW, Dr. Ch., Inorganic Salts Research Laboratory, Department of Chemistry, Bulgarian Academy of Sciences, "Acad. G. Bontchev-Street" Bl.11, Sofia 1040, Bulgaria
(TEL: 441551. TELEX: 22729 ECH BAN BG) (30)

BALL, Dr. W., BASF Aktiengesellschaft, Polyolefin Abteilung, Carl Bosch Strasse 38, D-6700 Ludwigshafen/Rhein, Federal Republic of Germany (25)

BALLSCHMITTER, Prof. K., Abteilung Analytische Chemie, Universität Ulm, Postfach 4066, Oberer Eselsberg, D-7900 Ulm, Federal Republic of Germany
(TEL: +49 (731) 1762181. TELEX: 0712567 UNIUL D) (27, 28)

BALOGH, Dr. I., Department of Forensic Medicine, Semmelweis University Medical School, Ulloi utca 93, H-1091 Budapest IX, Hungary (37)

BAMFORD, Prof. C. H., Bioengineering and Medical Physics Unit, Duncan Building, Faculty of Medicine, University of Liverpool, Royal Liverpool Hospital, Liverpool L7 8XP, UK
(TEL: +44 (51) 709 0141. TELEX: 627095 G) (10, 23, 26)

BANGHAM, Dr. D. R., National Institute for Biological Standards and Control, Holly Hill, Hampstead, London NW3 6RB, UK
(TEL: +44 (1) 435 2232. TELEX: 21911 NIBSAC LDN G) (36)

BANSAL, Prof. R. C., Chemistry Department, Panjab University, Chandigarh-160014, India
(TEL: 31435) (20)

BANTJES, Prof. A., Chemical Technology Department, Technische Hogeschool Twente, PO Box 217, 7500 AE Enschede, Netherlands (23)

BARBER, Prof. R. C., Department of Physics, University of Manitoba, Winnipeg, Manitoba R3T 2N2, Canada
(TEL: +1 (204) 474 9289. TELEX: 07-55264 CA) (19)

BĂRBULESCU, Prof. N., Institutul de Cercetări Chimice (ICECHIM), 202 Splaiul Independenţei, sect. 6, R-77208 Bucureşti, Romania
(TELEX: 10944 ICECHIM R) (34)

BAREISS, Dr. R. E., Die Makromolekulare Chemie, Hegelstrasse 45, D-6500 Mainz, Federal Republic of Germany
(TEL: +49 (6131) 381719) (23)

BARNES, Dr. I. L., A25 Physics Building, National Bureau of Standards, US Department of Commerce, Washington, DC 20234, USA
(TEL: +1 (301) 921 2781. TELEX: 898493 GARG) (19)

BARRETT, Dr. J. W., Cole Group PLC, Whitecliff House, 852 Brighton Road, Purley, Surrey CR2 2UY, UK
(TEL: +44 (1) 668 4161. TELEX: 23864 G) (10, 24)

BARTHEL, Prof. J. M. G., Institut für Physikalische und Theoretische Chemie, Universität Regensburg, Universitätsstrasse 31, D-8400 Regensburg, Federal Republic of Germany
(TEL: +49 (941) 943 4044/4045. TELEX: 065658 UNIRE D) (16)

BARTON, Prof. A. F. M., School of Mathematical and Physical Sciences, Murdoch University, Murdoch, Western Australia 6157, Australia
(TEL: +61 (9) 332 2211. TELEX: AA92711 MULIB) (30)

BARTON, Dr. J. M., Materials Department, Royal Aircraft Establishment, Farnborough, Hampshire GU14 6TD, UK (26)

BATES, Mr. J. A. R., Pesticide Registration Department, Ministry of Agriculture, Fisheries and Food, Harpenden Laboratory, Hatching Green, Harpenden, Hertfordshire AL5 2BD, UK
(TEL: +44 (5827) 5241) (33)

BATICH, Prof. C., Department of Materials Science and Engineering, 246 Rhines Hall, University of Florida, Gainesville, Florida 32611, USA
(TEL: +1 (904) 392 1461) (26)

BATLEY, Dr. G. E., Division of Energy Chemistry, Lukas Heights Research Laboratories, Commonwealth Scientific and Industrial Research Organization, Private Mail Bag 7, Sutherland, New South Wales 2232, Australia (29)

BATTAGLIA, Dr. R., Kantonales Labor Zürich, Fehrenstrasse 15, CH-8030 Zürich, Switzerland
(TEL: +41 (1) 252 56 54) (32)

BATTINO, Prof. R., Department of Chemistry, Wright State University, Dayton, Ohio 45435, USA
(TEL: +1 (513) 873 2477) (30)

BAUMARD, Dr. J. F., Centre de Recherches sur la Physique des Hautes Températures du Centre National de la Recherche Scientifique, 1D avenue de la Recherche Scientifique, F-45045 Orléans Cédex, France
(TEL: +33 (38) 63 06 17. TELEX: 760351 CNRSORL F) (20)

BAYER, Prof. G., Institut für Kristallographie und Petrographie der Eidgenössischen Technischen Hochschule Zürich, Sonneggstrasse 5, CH-8092 Zürich, Switzerland
(TEL: +41 (1) 256 37 35. TELEX: 53178 ETHBI CH) (20)

BAZAN, Dr. N. G., Avenida Alem 1253, 8000 Bahía Blanca, Argentina (33)

BEARE-ROGERS, Dr. JOYCE L., Department of National Health and Welfare, Food Directorate, Sir Frederick G. Banting Research Centre, Tunney's Pasture, Ottawa, Ontario K1A 0L2, Canada
(TEL: +1 (613) 593 4449. TELEX: 0533679 CA) (33)

BECK, Prof. M. T., Fizikai Kémia Tanszék, Kossuth Lajos Tudományegyetem, POB 7, Egyetem tér, H-4010 Debrecen X, Hungary
(TEL: +36 (52) 161 666. TELEX: 72200 UNIVK H) (19, 29, 31)

BECKER, Dr. E. D., Room 118, Building 1, National Institutes of Health, Bethesda, Maryland 20205, USA
(TEL: +1 (301) 496 2215) (15)

BECKER, Prof. G. W., Bundesanstalt für Materialprüfung (BAM), Unter den Eichen 87, D-1000 Berlin 45, Federal Republic of Germany
(TEL: +49 (30) 810410009. TELEX: 183261 BAMB D) (23, 24)

BÉDARD, Mr. M., Environmental Services, POB 415, Noranda, Province Québec J9S 5C4, Canada (37)

BEEZER, Dr. A. E., Department of Chemistry, Chelsea College, Manresa Road, London SW3 6LX, UK
(TEL: +44 (1) 351 2488) (13)

BEHNE, Dr. D., Hahn-Meitner-Institut, Glienickerstrasse 100, D-1000 Berlin 39, Federal Republic of Germany
(TEL: +49 (30) 800927484. TELEX: 0185763 D) (38)

BEHRENS, Prof. D., DECHEMA, Postfach 970146, Theodor-Heuss-Allee 25, D-6000 Frankfurt/Main 97, Federal Republic of Germany
(TEL: +49 (611) 756 4235. TELEX: 0412490 D) (5)

BELAICH, Prof. J. P., Laboratoire Chimie Bacterienne du Centre National de la Recherche Scientifique, 31 chemin Joseph-Aiguier, F-13274 Marseille Cédex 2, France
(TEL: +33 (91) 71 90 42. TELEX: 430 225 CNRSMAR F) (13)

BELETSKAYA, Prof. IRINE P., Faculty of Chemistry, Lomonosov Moscow State University, Leninskie Gory, SU-117234 Moscow, V-234, USSR
(TEL: +7 (095) 139 3919) (21)

BELLENGER, Mme. V., Ecole Nationale Supérieure des Arts et Metiers (ENSAM), 151 boulevard de l'Hôpital, F-75640 Paris Cédex 13, France
(TEL: +33 (1) 336 49 55. TELEX: 220064 ETRAV F, EXT. 379) (25)

BENISKA, Prof. J., Faculty of Chemical Technology, Vysoká Škola Chemicko-Technologická, Jánská 1, CS-880 37 Bratislava, Czechoslovakia (24)

BERESNIEWICZ, Dr. A., Central Research and Development Department, E. I. du Pont de Nemours and Co. Inc., Experimental Station, Building 328/202, Wilmington, Delaware 19898, USA
(TEL: +1 (302) 772 1800. TELEX: 835420 DUPONT) (26)

BERGE, Dr. A., Institutt for Industriell Kjemi/SINTEF, Universitet i Trondheim-Norges Tekniske Høgskole, N-7034 Trondheim-NTH, Norway (25)

BERGER, Dr. K., Institut für Physikalische Chemie, Johannes-Gutenberg Universität Mainz, D-6500 Mainz, Federal Republic of Germany
(TEL: +49 (6131) 392330. TELEX: 4187476 UNI D) (26)

BERLIN, Prof. A., Commission des Communautés Européennes, Direction Générale de l'Emploi des Affaires Sociales et de l'Education, Bâtiment Jean Monnet, Platêau du Kirchberg, Luxembourg
(TEL: 43011) (38)

BERMAN, Mrs. E., Division of Biochemistry, Cook County Hospital, 1825 West Harrison, Chicago, Illinois 60614, USA (37)

BERTELLO, Prof. L. F., Perú 420, 1640 Martínez, Buenos Aires, Argentina
(TEL: +54 (1) 7922128) (19, 36)

BEVINGTON, Prof. J. C., Department of Chemistry, University of Lancaster, Bailrigg, Lancaster LA1 4YA, UK
(TEL: +44 (524) 65201) (26)

BEWICK, Dr. A., Department of Chemistry, University of Southampton, Highfield, Southampton SO9 5NH, UK (16)

BEYERMANN, Prof. K., Institut für Anorganische und Analytische Chemie, Johannes-Gutenberg Universität Mainz, Postfach 3980, Saarstrasse 21, D-6500 Mainz, Federal Republic of Germany
(TEL: +49 (6131) 392101. TELEX: 4187476 UNI D) (27, 28)

BEYNON, Prof. J. H., Department of Chemistry, University College of Wales, Singleton Park, Swansea SA2 8PP, UK
(TEL: +44 (792) 205678) (17)

BHARDWAJ, Dr. I. S., Indian Petrochemicals Corpn. Ltd., PO Petrochemicals, Dist. Baroda 391346, Gujarat, India
(TEL: 7411/7611/7661. TELEX: 0175-364, 0175-365 IPCL IN) (24)

BIEMANN, Prof. K., Department of Chemistry, Room 56-010, Massachusetts Institute of Technology, Cambridge, Massachusetts 02139, USA
(TEL: +1 (617) 253 7221. TELEX: 92 1473 MITCAM) (17, 27, 28)

BIERENS DE HAAN, Dr. J., Laboratoire Riotton SA, 16 boulevard des Tranchées, CH-1206 Genève, Switzerland
(TEL: +41 (22) 46 89 33. TELEX: 28190 RIO CH) (35)

BIETHAN, Dr. U., Chemische Werke Hüls AG, Postfach 1320, D-4370 Marl-Hüls 1, Federal Republic of Germany
(TELEX: 829211-0 CW D) (25)

BIKALES, Dr. N. M., Polymer Program, National Science Foundation, Washington, DC 20550, USA
(TEL: +1 (202) 357 9789. TELEX: 89-2438) (23)

BILLEK, Prof. G., Unilever Forschungsgesellschaft mbH, Postfach 1568, Behringstrasse 154, D-2000 Hamburg 50, Federal Republic of Germany
(TEL: +49 (40) 8827702. TELEX: 2164293 UNAT D) (32)

BJÖRKMAN, Prof. A., Instituttet for Kemiindustri, Danmarks Tekniske Højskole, Bygning 227, DK-2800 Lyngby, Denmark
(TEL: +45 (2) 88 41 44. TELEX: 37529 DTHDIA DK) (v, 10, 12)

BLACKWOOD, Dr. J. E., Chemical Abstracts Service, POB 3012, 2540 Olentangy River Road, Columbus, Ohio 43210, USA
(TELEX: 810 482 1608 CHEM ABS COL) (22)

BLÁHA, Dr. K., Ústav Organické Chemie a Biochemie, Československá Akademie Věd, Flemingovo náměstí 2, CS-166 10 Praha 6-Dejvice, Czechoslovakia
(TEL: +42 (2) 320141) (21)

BLAIR, Dr. E., Health and Environmental Sciences, Dow Chemical Co., 2020 Dow Center, Midland, Michigan 48640, USA (10)

BLOCK, Prof. J. H., Abteilung Grenzflächenreaktionen, Fritz-Haber-Institut der Max-Planck-Gesellschaft, Faradayweg 4-6, D-1000 Berlin 33, Federal Republic of Germany
(TEL: +49 (30) 8305411. TELEX: 185676 FHIMP D) (6, 17, 18)

DEN BOEF, Prof. G., Laboratorium voor Analytische Scheikunde, J. H. van't Hoff Instituut, Universiteit van Amsterdam, Nieuwe Achtergracht 166, 1018 WV Amsterdam, Netherlands
(TEL: +31 (20) 5223515. TELEX: 16460 ANAL CHEM NL) (11, 27)

BOEHM, Prof. H. P., Institut für Anorganische Chemie der Universität München, Meiserstrasse 1, D-8000 München 2, Federal Republic of Germany
(TEL: +49 (89) 5902355) (20)

BOHNSACK, Dr. G., Bayer AG, D-5090 Leverkusen-Bayerwerk, Federal Republic of Germany
(TELEX: 85103 0 BY D) (31)

BOHUON, Prof. C., Institut Gustave Roussy, rue Camille Desmoulins, F-94800 Villejuif, France
(TEL: +33 (1) 559 49 09) (36)

DE BOLSTER, Dr. M. W. G., Vakgroep Anorganische en Theoretische Chemie, Subfaculteit der Scheikunde, Vrije Universiteit, De Boelelaan 1083, 1081 HV Amsterdam, Netherlands
(TEL: +31 (20) 5485738. TELEX: 11329 NL) (19)

DE BONI, Dr. U., Department of Physiology, University of Toronto, Toronto, Ontario M5S 1A8, Canada
(TEL: +1 (416) 978 2682) (38)

BONINI, Prof. P., Laboratorio Analisi, Ospedale San Raffaele, Via Olgettina 60, I-20132 Milano, Italy
(TEL: +39 (2) 21702315. TELEX: 313627 GEXPO I, ATTN. PROF. BONINI) (35)

BONNETAIN, Prof. L., Laboratoire d'Adsorption et Reactions des Solids, Institut National Polytechnique de Grenoble, ERA CNRS 368, Domaine Universitaire, BP 44, F-38401 Saint Martin d'Hères, France
(TEL: +33 (76) 54 41 27) (20)

BORSY, Dr. J., Gyógyszerkutató Intézet, Szabadságharcosok utca 47-49, H-1045 Budapest, Hungary
(TEL: +36 (1) 691425. TELEX: 224219 H) (36)

BORZANI, Prof. W., Instituto de Pesquisas Tecnológicas do Estado de São Paulo SA - IPT, Divisao de Química e Engenharia Química, Caixa Postal 7141, 05508 São Paulo - SP, Brazil
(TEL: +55 (11) 2682211. TELEX: (011) 22831-INPT BR) (32)

BOS, Prof. H. J. T., Organisch Chemisch Laboratorium van de Rijksuniversiteit, Croesestraat 79, 3522 AD Utrecht, Netherlands (21)

BOULOS, Prof. M., Department of Chemical Engineering, University of Sherbrooke, Sherbrooke, Québec J1K 2R1, Canada
(TEL: +1 (819) 565 4317) (15)

BOULTON, Dr. A. J., School of Chemical Sciences, University of East Anglia, University Plain, Norwich NR4 7TJ, UK
(TEL: +44 (603) 56161. TELEX: 97154 G) (21)

BOURG, Dr. A. C. M., Département Minéralogie, Géochimie, Analyses, Bureau de Recherches Géologiques et Minières, BP 6009, F-45060 Orléans Cédex, France
(TEL: +33 (38) 63 80 01. TELEX: 780258 BRGMA F) (29)

BOUSQUET, Prof. B., Laboratoire Central de Biochimie, Hôpital Saint-Louis, 2 place du Dr. Albert Fournier, F-75475 Paris Cédex 10, France
(TEL: +33 (1) 206 33 12) (36)

BOWMAKER, Dr. G. A., Department of Chemistry, University of Auckland, Private Bag, Auckland, New Zealand
(TEL: +64 (9) 792300. TELEX: 21480 NZ) (30)

BOYER, Prof. R. F., Midland Molecular Institute, 1910 West St. Andrews, Midland, Michigan 48640, USA (26)

BOYSEN, Dr. M., Department of Pathology, Norsk Hydros Institute for Cancer Research, Norwegian Radium Hospital, Montebello, Oslo 3, Norway
(TEL: +47 (2) 554080) (37)

BOZEC, Mr. Ch., Département de Direction Technique, Societé Metallurgique le Nickel, 33 avenue du Maine, F-75755 Paris Cédex 15, France
(TEL: +33 (1) 538 48 48) (37)

BRACHA, Dr. P., Machteshim Ltd., Beer-Sheva, Israel (34)

BRADSHAW, Prof. A. M., Fritz-Haber-Institut der Max-Planck-Gesellschaft, Faradayweg 4-6, D-1000 Berlin 33, Federal Republic of Germany
(TEL: +49 (30) 8305501. TELEX: 185676 FHIMP D) (17)

BRASLAVSKY, Prof. SILVIA E., Max-Planck-Institut für Strahlenchemie, Stiftstrasse 34-36, D-4330 Mülheim/Ruhr 1, Federal Republic of Germany
(TEL: +49 (208) 31073. TELEX: 856741 MPSTR D) (22)

BRAUN, Prof. D., Deutsches Kunststoff-Institut, Schlossgartenstrasse 6 R, D-6100 Darmstadt, Federal Republic of Germany
(TEL: +49 (6151) 162104) (24, 25)

BREMER, Prof. H., Technische Hochschule "Carl Schorlemmer" Leuna-Merseburg, Geusaer Strasse, DDR-4200 Merseburg 6, German Democratic Republic
(TEL: +37 (442) 462101. TELEX: 471320 THLM DD) (11)

BREUER, Prof. J., Zentrallaboratorium des Marienhospitals, Virchowstrasse 135, D-4650 Gelsenkirchen, Federal Republic of Germany (35)

BRINGI, Dr. N. V., Hindustan Lever Research Centre, Chakala, Andheri (East), Bombay-400099, India
(TEL: +91 (22) 6329471) (33)

BRINK, Dr. A., SASOL Ltd., PO Box 1, Sasolburg 9570, Republic of South Africa (10)

BRODERSEN, Prof. S., Kemisk Institut, University of Aarhus, Langelandsgade 140, DK-8000 Aarhus C, Denmark
(TEL: +45 (6) 12 46 33) (17)

BROWN, Prof. R. D., Department of Chemistry, Monash University, Wellington Road, Clayton, Victoria 3168, Australia
(TEL: +61 (3) 541 3550. TELEX: AA32691 MONASH) (17)

BROWN, Dr. S. S., Regional Laboratory for Toxicology, Dudley Road Hospital, POB 293, Birmingham B18 7QH, UK
(TEL: +44 (21) 554 3801) (35, 37)

BRUNNER, Dr. E., BASF Aktiengesellschaft, Ammoniaklaboratorium WAA/M325, D-6700 Ludwigshafen/Rhein, Federal Republic of Germany
(TEL: +49 (621) 604559. TELEX: 464680 BASF D) (15)

BRÜSCHWEILER, Dr. H., Eidgenössische Materialprüfungs- und Versuchsanstalt für Industrie, Bauwesen und Gewerbe, Unterstrasse 11, CH-9001 St. Gallen, Switzerland
(TEL: +41 (71) 20 91 71. TELEX: 71278 EMPA CH) (33)

BRYSON, Dr. D. D., Imperial Chemical Industries PLC, Agricultural Division, POB 8, Billingham, Cleveland TS23 1LE, UK (37)

BUCAT, Dr. R. B., Department of Physical and Inorganic Chemistry, University of Western Australia, Nedlands, Western Australia 6009, Australia
(TEL: +61 (9) 380 3158. TELEX: AA92992) (11)

BÜCHEL, Prof. K. H., Bayer AG, D-5090 Leverkusen-Bayerwerk, Federal Republic of Germany
(TEL: +49 (214) 30 71375. TELEX: 85103 0 BY D) (10)

BUCKNALL, Dr. C. B., Cranfield Institute of Technology, School of Industrial Science, Cranfield, Bedford MK43 AOL, UK
(TEL: +44 (234) 750 111. TELEX: 825072 CRANTECH G) (24)

BUDEVSKI, Prof. E., Central Laboratory of Electrochemical Power Sources, Bulgarian Academy of Sciences, Sofia 1000, Bulgaria
(TEL: 723454) (16)

BUFFLE, Dr. J., Département de Chimie Minérale Analytique et Appliquée de l'Université, Section Chimie - Sciences II, 30 quai Ernest-Ansermet, CH-1211 Genève 4, Switzerland
(TEL: +41 (22) 21 93 55) (29)

BULL, Dr. J. R., National Chemical Research Laboratory, Council for Scientific and Industrial Research, PO Box 395, Pretoria 0001, Republic of South Africa
(TEL: +27 (12) 869211. TELEX: 3-630 SA) (21)

BULLOCK, Dr. KATHRYN R., Johnson Controls Inc., POB 591, 5757 North Green Bay Avenue, Milwaukee, Wisconsin 53072, USA (31)

BUNKER, Dr. P. R., Herzberg Institute of Astrophysics, National Research Council of Canada, Ottawa, Ontario K1A OR6, Canada
(TEL: +1 (613) 992 6917. TELEX: 053-3715 NRC OTT CA) (17)

BUNNETT, Prof. J. F., Division of Natural Sciences-II, University of California, Santa Cruz, California 95064, USA
(TEL: +1 (408) 429 2261) (21)

BURATTI, Dr. A., Istituto di Medicina del Lavoro, Via S. Barnaba 8, I-20122 Milano, Italy
(TEL: +39 (2) 581610/576091) (37)

BURTIS, Dr. C. A., Chemical Technology Division, Oak Ridge National Laboratory, POB X, Oak Ridge, Tennessee 37830, USA
(TEL: +1 (615) 576 2917. TELEX: 81572176 USDOE OKRE) (35, 36)

BUSCH, Prof. D. H., Department of Chemistry, Ohio State University, 140 West Eighteenth Avenue, Columbus, Ohio 43210, USA
(TEL: +1 (614) 422 6120) (19, 22)

BUSCHBECK, Dr. K. Ch., Gmelin-Institut für Anorganische Chemie und Grenzgebiete der Max-Planck-Gesellschaft, Carl Bosch Haus, Postfach 900467, Varrentrappstrasse 40-42, D-6000 Frankfurt/Main 90, Federal Republic of Germany
(TEL: +49 (611) 791 7215. TELEX: 412526 GMELI D) (19)

BUTLER, Dr. L. R. P., Information and Research Services, Council for Scientific and Industrial Research, PO Box 395, Pretoria 0001, Republic of South Africa
(TEL: +27 (12) 869211. TELEX: 3-630 SA) (28)

BÜTTNER, Prof. J., Institut für Klinische Chemie, Medizinische Hochschule Hannover, Karl-Wiechert-Allee 9, D-3000 Hannover 61, Federal Republic of Germany
(TEL: +49 (511) 5321) (35, 36)

BYLICKI, Prof. A., Instytut Chemii Fizycznej, Polskiej Akademii Nauk, ul. Kasprzaka 44-52, PL-01 224 Warszawa, Poland
(TELEX: 817097 ICHF PL) (29, 30)

CAIS, Prof. M., Department of Chemistry, Technion-Israel Institute of Technology, Technion City, Haifa 32000, Israel
(TEL: +972 (4) 225111. TELEX: 46650 TECLI IL) (35)

CALADO, Prof. J. C. G., Complexo I, Instituto Superior Técnico, P-1096 Lisboa, Portugal
(TEL: +351 (19) 572616) (16)

CAMPBELL, Prof. A. D., Department of Chemistry, University of Otago, POB 56, Dunedin, New Zealand
(TEL: +64 (24) 771640) (10, 27)

CAMPBELL, Dr. D. J., Vancouver General Hospital, Division of Clinical Chemistry, 855 West 12th Avenue, Vancouver, British Columbia V5Z 1M9, Canada
(TEL: +1 (604) 875 4130) (36)

CAMPBELL, Prof. P. N., Courtauld Institute of Biochemistry, Middlesex Hospital, Medical School, Mortimer Street, London W1P 7PN, UK
(TEL: +44 (1) 636 8333) (35)

CANÈ, Mr. F., Anic SpA (LAB/ATS), I-20097 S. Donato Milanese (Mi.), Italy (25)

CANN, Mr. J. M., Polymer Science, R&D Department, BP Chemicals PLC, Sully, Penarth, South Glamorgan CF6 2YU, UK
(TEL: +44 (446) 73 2321. TELEX: 498276 G) (24)

CANNON, Prof. J. R., Department of Organic Chemistry, University of Western Australia, Nedlands, Western Australia 6009, Australia
(TEL: +61 (9) 380 3152. TELEX: AA92992) (21)

CANTAFORA, Dr. A., Istituto Superiore di Sanità, Viale Regina Elena 299, I-00161 Roma, Italy
(TEL: +39 (6) 4990. TELEX: 71 ISTISAN RMO 071 I) (32)

CAPPACIOLI, Dr. J. H., Solis 1089 - 9° Piso, 1078 Buenos Aires, Argentina (28)

CARACULACU, Dr., Institute of Macromolecular Chemistry P. Poni, Aleia Ghica Voda 41 A, 600 Iasi, Romania (TEL: 41901) (25)

CARGILL, Dr. R. W., Department of Molecular and Life Sciences, Dundee College of Technology, Dundee DD1 1HG, UK (30)

CARLING, Dr. R. W., Sandia National Laboratories, Livermore, California 94550, USA (31)

CARLSON, Dr. D. J., Textile Chemistry Section, National Research Council of Canada, Division of Chemistry, Ottawa, Ontario K1A 0R6, Canada (26)

CARR, Mr. G. D., PO Box 50235, Musgrave Road, Natal 4062, Republic of South Africa (33)

CARREGA, Dr. M. E., Direction des Recherches, Rhône-Poulenc Spécialités Chimiques, F-92097 Paris la Defense Cédex 29, France
(TEL: +33 (1) 768 26 52. TELEX: 610500 RHÔNE F) (23)

CARTER, Dr. CYNTHIA G., NDAB - National Research Council, 2101 Constitution Avenue NW, Washington, DC 20418, USA (30)

CASTEGNARO, Prof. M. J. J., Unit of Environmental Carcinogens, International Agency for Research on Cancer, 150 cours Albert-Thomas, F-69372 Lyon Cédex, France
(TEL: +33 (7) 875 81 81. TELEX: 380023 F) (32)

CECCHINI, Prof. M. A. G., Departamento de Química do Instituto Tecnológico de Aeronáutica/CTA, 12.200 São José dos Campos - SP, Brazil
(TEL: +55 (123) 21 9088. TELEX: 01133393 BR, ATTN. ITA/CTA) (15)

CECUTTI, Dr. A. G., Medical Department, Falconbridge Nickel Mines Ltd., Sudbury Operations, Falconbridge, Ontario POM ISO, Canada
(TEL: +1 (705) 674 5530) (37)

CEDERGREN, Prof. A., Department of Analytical Chemistry, University of Umeå, S-901 87 Umeå, Sweden
(TEL: +46 (90) 165151) (28)

DE CEDIEL, Dr. NELLY, Ass. Columbiana de Especialistas en Laboratorio Clinico, 15-47 Calle 49, Bogota, Colombia
(TEL: 2-323681) (36)

ČEFELÍN, Dr. P., Ústav Makromolekulární Chemie, Československá Akademie Věd, Heyrovského náměstí 2, CS-162 06 Praha 616, Czechoslovakia
(TEL: +42 (2) 360341. TELEX: 122019 IMCP C) (23)

CHALLA, Prof. G., Laboratorium voor Polymeerchemie RUG, Nyenborgh 16, 9747 AG Gröningen, Netherlands
(TEL: +31 (50) 117215. TELEX: 53410 RUGRO NL) (23)

CHAN, Prof. C. Y., Department of Chemistry, University of Malaya, Kuala Lumpur, Malaysia (31)

CHAN, Prof. K. C., Jabatan Kimia, Universiti Malaya, Lembah Pantai, Kuala Lumpur 22-11, Malaysia
(TEL: +60 (3) 555466. TELEX: 37453 MA) (21)

CHANG, Prof. T. L., Department of Chemistry, Peking University, Beijing, China
(TEL: 282471. TELEX: 222 39 PKUNI CN) (19)

CHASTRETTE, Prof. M., Laboratoire de Chimie Organique Physique, Université Claude-Bernard LYON I, 43 boulevard du 11 Novembre 1918, F-69622 Villeurbanne Cédex, France
(TEL: +33 (7) 889 81 24. TELEX: 42009 F) (11)

CHIAVARINI, Dr. M., Centro Ricerche, Montedison SpA/DIPI, Stabilimento di Castellanza, Via Sempione 13, I-21053 Castellanza, Italy (25)

CHIU, Dr. J., Polymer Products, Central Research and Development Department, E. I. du Pont de Nemours and Co. Inc., Experimental Station, Wilmington, Delaware 19898, USA
(TELEX: 835420 DUPONT) (26)

CHOBANOV, Dr. D. G., Institute of Organic Chemistry, Centre of Phytochemistry, Bulgarian Academy of Sciences, Sofia 1113, Bulgaria
(TEL: 7131. TELEX: 22729 ECH BAN BG) (33)

CHOPPIN, Prof. G. R., Department of Chemistry, Florida State University, Tallahassee, Florida 32306, USA
(TEL: +1 (904) 644 3875) (29)

CHRISTENSEN, Mr. G., A/S Sadolin og Holmblad, Holmbladsgade 70, DK-2300 København S, Denmark (25)

CHRISTENSEN, Dr. J. M., Danish National Institute of Occupational Health, Baunegaardsvej 73, DK-2900 Hellerup (Copenhagen), Denmark
(TELEX: 16149 AMI DK) (37)

CHRISTENSEN, Mrs. J. M., Danish National Institute of Occupational Health, Baunegaardsvej 73, DK-2900 Hellerup (Copenhagen), Denmark
(TEL: +45 (1) 68 28 68. TELEX: 16149 AMI DK) (37)

CHRISTOPHERSEN, Dr. B., Klinisk-Kjemisk Avdeling, Rikshospitalet, Oslo 1, Norway (36)

CHURAEV, Prof. N. V., Institute of Physical Chemistry, Academy of Sciences of USSR, Leninskii Prospekt 31, SU-117312 Moscow, B-312, USSR
(TEL: +7 (095) 234 0014) (17)

CLAIBORNE IRBY, Dr. E., Reynolds Metal Co., Executive Office Building, Richmond, Virginia 23261, USA
(TEL: +1 (804) 281 2773) (38)

CLARK, Dr. J. B., National Institute for Materials Research, Council for Scientific and Industrial Research, PO Box 395, Pretoria 0001, Republic of South Africa
(TEL: +27 (12) 869211. TELEX: 3-630 SA) (20)

CLARK, Prof. D. T., Department of Chemistry, University of Durham, Science Laboratories, South Road, Durham DH1 3LE, UK
(TEL: +44 (385) 64971) (25)

CLEGG, Dr. P. L., 1 High Oaks Road, Welwyn Garden City, Hertfordshire AL8 7BJ, UK
(TEL: +44 (7073) 24678) (23, 24)

CLEVER, Prof. H. L., Department of Chemistry, Emory University, Atlanta, Georgia 30322, USA
(TEL: +1 (404) 329 6616) (30)

COETZEE, Prof. J. F., Department of Chemistry, University of Pittsburgh, Pittsburgh, Pennsylvania 15260, USA
(TEL: +1 (412) 624 5090. TELEX: 812 466) (27)

COHEN, Dr. E. R., Science Center, Rockwell International, 1049 Camino dos Rios, Thousand Oaks, California 91360, USA
(TEL: +1 (805) 498 4545) (15)

COHEN-ADAD, Dr. MARIE-THERESE, c/o Prof. R. Cohen-Adad, Laboratoire de Physicochimie, Université Claude Bernard LYON I, 43 boulevard du 11 Novembre 1918, F-69622 Villeurbanne Cédex, France
(TEL: +33 (7) 889 80 60. TELEX: 420009 F) (31)

COHEN-ADAD, Prof. R., Laboratoire de Physicochimie, Université Claude Bernard LYON I, 43 boulevard du 11 Novembre 1918, F-69622 Villeurbanne Cédex, France
(TEL: +33 (7) 889 80 60. TELEX: 420009 F) (30)

COLE, Prof. A. R. H., Department of Physical and Inorganic Chemistry, University of Western Australia, Nedlands, Western Australia 6009, Australia
(TEL: +61 (9) 380 3162. TELEX: AA92992) (17)

COLES, Dr. L. E., Mid-Glamorgan Public Health Laboratory, The Parade, Cardiff CF2 3UJ, UK
(TEL: +44 (222) 28033. TELEX: 27950 REF. NO. 695 MONREF. G) (11, 32)

COLIN, Prof. R., Laboratoire de Chimie Physique Moléculaire, Université Libre de Bruxelles, 50 Avenue F. D. Roosevelt, B-1050 Bruxelles, Belgium
(TEL: +32 (2) 649 00 30. TELEX: 23069 UNILIB BRUX B) (17)

COMETTO, Dr. C., Centro Ricerche "Giulo Natta", Montedison SpA, I-44100 Ferrara, Italy
(TEL: +39 (532) 51251/51711. TELEX: 310679 MONTED I) (25)

CONNELLY, Dr. J. C., Department of Biochemistry, Royal Children's Hospital, Flemington Road, Parkville, Victoria, Australia
(TEL: +61 (3) 345 5522. TELEX: AA37281) (35)

CONNOR, Prof. J. A., Department of Chemistry, University of Kent at Canterbury, Canterbury, Kent CT2 7NH, UK
(TEL: +44 (227) 66822. TELEX: 965449 G) (19)

CONSTANTIN, Dr. D., Societé Chimique des Charbonnages, CdF Chimie, Usines du Nord, BP 1, F-62160 Bully les Mines, France
(TEL: +33 (21) 79 92 33. TELEX: 120142 F) (24, 25)

CONWAY, Prof. B. E., Department of Chemistry, University of Ottawa, Ottawa, Ontario K1N 6N5, Canada
(TEL: +1 (613) 231 2431) (16)

COONEY, Prof. C. L., Department of Nutrition and Food Science, Room 66-468, Massachusetts Institute of Technology, Cambridge, Massachusetts 02139, USA
(TEL: +1 (617) 253 3108. TELEX: 92 1473 MITCAM) (32)

COPLEN, Dr. T. B., United States Geological Survey, 431 National Center, 12201 Sunrise Valley Drive, Reston, Virginia 22092, USA
(TEL: +1 (703) 860 6083. TELEX: 89-9153) (19)

CORISH, Prof. J., Chemistry Department, Trinity College, University of Dublin, Dublin 2, Ireland
(TEL: +0044 (0001) 772941. TELEX: 25442 TCD EI, ATTN. CORISH) (20)

CORNISH-BOWDEN, Dr. A., Department of Biochemistry, University of Birmingham, POB 363, Birmingham B15 2TT, UK
(TEL: +44 (21) 472 1301. TELEX: 338938 SPAPHY G, ATTN. CORNISH-BOWDEN, BIOCHEMISTRY) (13)

CORTÉS, Dr. L., Escuela de Química, Universidad Central de Venezuela, Caracas, Venezuela (12)

COSTA, Dr. M., Department of Pharmacology, Division of Toxicology, University of Texas Medical School, POB 20708, Houston, Texas 77025, USA
(TEL: +1 (731) 742 5977) (37)

COSTER, Mr. H., Borg-Warner Chemicals, PO Box 8122, Amsterdam, Netherlands
(TEL: +31 (20) 139833. TELEX: 16257 NL) (24)

COUNIOUX, Prof. J. J., c/o Prof. R. Cohen-Adad, Laboratoire de Physicochimie, Université Claude Bernard LYON I, 43 boulevard du 11 Novembre 1918, F-69622 Villeurbanne Cédex, France
(TEL: +33 (7) 889 80 60. TELEX: 420009 F) (31)

COUTURES, Dr. J.-P., Centre de Recherches sur la Physique des Hautes Températures du Centre National de la Recherche Scientifique, 1D avenue de la Recherche Scientifique, F-45045 Orléans Cédex, France
(TEL: +33 (38) 63 06 17. TELEX: 760351 CNRSORL F) (20)

COVINGTON, Dr. A. K., Department of Physical Chemistry, School of Chemistry, Bedson Building, University of Newcastle upon Tyne, Newcastle upon Tyne NE1 7RU, UK
(TEL: +44 (632) 328511. TELEX: 53654 UNINEW G) (16, 29)

COWIE, Prof. J. M. G., Department of Chemistry, University of Stirling, Stirling FK9 4LA, UK
(TEL: +44 (786) 3171) (24)

COYLE, Dr. T. D., Inorganic Materials Division, National Bureau of Standards, US Department of Commerce, Washington, DC 20234, USA
(TEL: +1 (301) 921 2847. TELEX: 898493 GARG) (12, 19)

CRAMERS, Prof. C. A. M. G., Laboratorium voor Instrumentele Analyse, Technische Hogeschool Eindhoven, Postbus 513, 5600 MB Eindhoven, Netherlands
(TEL: +31 (40) 473026. TELEX: 51163 NL) (28)

CROISSANT, Dr. P., GAMS, 88 boulevard Malesherbes, F-75008 Paris, France
(TEL: +33 (1) 563 93 04) (5)

CROVINI, Prof. L., Istituto di Metrologia "Gustavo Colonnetti" del Consiglio Nazionale delle Ricerche, Strada delle Cacce 73, I-10135 Torino, Italy
(TEL: +39 (11) 3487834. TELEX: 212209 I) (15)

CUNNINGHAM, Dr. G. P., PPG Industries Inc., POB 9, Rosanna Drive, Allison Park, Pennsylvania 15101, USA (25)

CURRIE, Dr. L. A., Center for Analytical Chemistry, National Bureau of Standards, US Department of Commerce, Washington, DC 20234, USA
(TEL: +1 (301) 921 2440. TELEX: 898493 GARG) (28)

CURTIUS, Prof. A. J., Departamento Química - PUC, R. Marquês S. Vicente 225, 22451 Rio de Janeiro - RJ, Brazil (28)

CZEDIK-EYSENBERG, Dr. P. B., Österreichische Unilever GmbH, Schenkenstrasse 8, A-1011 Wien, Austria
(TEL: +43 (222) 66190. TELEX: 114080 UNIWN A) (32)

DAAE, Dr. L., Sentrallaboratoriet, Ullevål Sykehus, Oslo 1, Norway (37)

D'AGOSTINO, Prof. R., Instituto di Chimica Generale, Università di Bari, I-70126 Bari, Italy
(TEL: +39 (80) 330044) (15)

DAMS, Prof. R., Institute of Nuclear Sciences, Proeftuinstraat 86, B-9000 Gent, Belgium
(TEL: +32 (91) 22 87 21. TELEX: 12754 RUGENT B) (27)

DANTINNE, Dr. P., Laboratoire Central, Solvay & Cie SA, 310 Rue de Ransbeek, B-1120 Bruxelles, Belgium
(TELEX: 26619 SOLVAY NOH BT B) (26)

DAS, Prof. H. A., Energie Onderzoek Centrum Nederland, Postbus 1, 1755 ZG Petten, Netherlands
(TEL: +31 (2246) 6262) (38)

D'ASCENZO, Prof. G., Dipartimento Scienze Chimiche, Università di Camerino, Via S. Agostino 1, I-62032 Camerino, Italy (35)

DAVIDEK, Prof. J., Děkanát Fakulty Potravinářské a Biochemické Technologie, Vysoká Škola Chemicko-Technologická v Praze, Suchbátarova 5, CS-166 28 Praha 6-Dejvice, Czechoslovakia (32)

DAVISON, Dr. W., Freshwater Biological Association, Windermere Laboratory, The Ferry House, Ambleside, Cumbria LA22 OLP, UK
(TEL: +44 (9662) 2468/9) (29, 31)

DAY, Dr. J. P., Department of Chemistry, University of Manchester Institute of Science and Technology, Manchester M60 1QD, UK
(TELEX: 666094 G) (38)

DEÁK, Prof. G., Magyar Tudományos Akadémia, Kisérleti Orvostudományi Kutató Intézete, Szigony utca 43, H-1083 Budapest, Hungary
(TEL: +36 (1) 134 458) (21)

DE BIÈVRE, Prof. P., Central Bureau for Nuclear Measurements, Commission of the European Communities, Steenweg op Retie, B-2440 Geel, Belgium
(TEL: +32 (14) 57 16 05. TELEX: 33589 EURAT B) (19, 37)

DE BRUILLE, Dr. R., Laboratoire Central, Solvay & Cie SA, 310 Rue de Ransbeek, B-1120 Bruxelles, Belgium
(TEL: + 32 (2) 267 38 30. TELEX: 26619 SOLVAY NOH BT B) (25)

DEGAIN, Dr. R., Rhône-Poulenc SA, 25 quai Paul-Doumer, F-92408 Courbevoie Cédex, France (10, 11)

DELBRÜCK, Prof. A. F., Institut für Klinische Chemie II im Zentrum Laboratoriumsmedizin, Medizinische Hochschule Hannover, Podbielskiestrasse 380, D-3000 Hannover 51, Federal Republic of Germany
(TEL: +49 (511) 6461468) (36)

DELLWEG, Prof. H., Institut für Gärungsgewerbe und Biotechnologie, Seestrasse 13, D-1000 Berlin 65, Federal Republic of Germany
(TEL: +49 (30) 4509218. TELEX: 184 403 VLSF D) (32)

DELHAYE, Prof. M., Laboratoire de Chimie Physique du Centre National de la Recherche Scientifique, 2-8 rue Henri Dunant, F-94320 Thiais, France (17)

DELVES, Dr. H. T., Trace Metal Unit, Department of Chemical Pathology and Human Metabolism, South Laboratory and Pathology Block, Level D, Southampton General Hospital, Tremona Road, Southampton SO9 4XY, UK (37, 38)

DE MARIA, Prof. G., Centro di Termodinamica Chimica alle alte Temperature - CNR, c/o Istituto di Chimica Fisica, Università di Roma, I-00185 Roma, Italy
(TEL: +39 (6) 492950) (20)

DE NEVE, Dr. R., Sampling and Assays Laboratories, N. V. Metallurgie Hoboken-Overpelt, A. Greinerstraat 14, B-2710 Hoboken, Belgium (37)

DENISOV, Prof. E. T., Branch of the Institute of Chemical Physics, Academy of Sciences of USSR, Chernogolovka, Noginsk District, Moscow Region, 142432, USSR
(TELEX: 205792 ATOM SU) (15)

DERENDJAEV, Dr. B. G., Department of Molecular Spectroscopy, Institute of Organic Chemistry, Siberian Branch of Academy of Sciences of USSR, Prospekt Nauki 17, SU-630090 Novosibirsk 90, USSR
(TEL: +7 (3832) 651 660) (17)

DE SCHRYVER, Prof. F. C., Afdeling Organische Scheikunde, Katholieke Universiteit te Leuven, Celestijnenlaan 200 F, B-3030 Heverlee, Belgium
(TEL: +32 (16) 20 0656. TELEX: 23674 KULEUV B) (5, 22)

DESHPANDE, Prof. D. D., Department of Chemistry, Indian Institute of Technology, Powai, Bombay-400076, India
(TEL: +91 (22) 581421. TELEX: 011-71385 IITB IN) (23)

DESMORAS, Dr. J., Centre Nicolas Grillet, Rhône-Poulenc Industries, 13 quai Jules Guesde, F-94400 Vitry-sur-Seine, France
(TEL: +33 (1) 680 85 11. TELEX: 270064 RHÔNE F) (33)

DEUTSCH, Dr. SARAH, Laboratoires de Minéralogie et de Pétrologie, Université Libre de Bruxelles, 50 Avenue F. D. Roosevelt, B-1050 Bruxelles, Belgium
(TEL: +32 (2) 649 00 30. TELEX: 23069 UNILIB BRUX B) (6)

DEWALLY, Dr. D., Bâtiment C6, Chimie Minerale Appliquée, Université des Sciences et Techniques de Lille, F-59655 Villeneuve d'Ascq Cédex, France
(TEL: +33 (20) 91 92 22) (37)

DIEFENDORF, Prof. R. J., Material Engineering Department, School of Engineering, Rensselear Polytechnic Institute, Troy, New York 12181, USA
(TEL: +1 (518) 270 6371) (20)

DIEFFENBACHER, Dr. A., Research Department, Nestlé Products Technical Assistance Co. Ltd., Case Postale 88, CH-1814 La Tour-de-Peilz, Switzerland
(TEL: +41 (21) 51 02 11. TELEX: 451333 NTA CH) (33)

DIELEMAN, Dr. J., Philips Research Laboratories, Prof. Holst Laan, 5600 MB Eindhoven, Netherlands (15)

VAN DIJK, Dr., Plastics and Rubber Research Institute TNO, PO Box 71, 2600 AB Delft, Netherlands
(TEL: +31 (15) 569330. TELEX: 31453 ZP TNO NL) (24)

DILARIS, Prof. IRENE, Laboratory of Organic Chemistry, University of Athens, Navarinou Str. 13a, Athens, Greece
(TEL: +30 (1) 3636211) (11)

DILKS, Dr. A., Webster Research Center, Xerox Corpn., 800 Phillips Road, Webster, New York 14580, USA (26)

DIRKSE, Prof. T. P., Department of Chemistry, Calvin College, Grand Rapids, Michigan 49506, USA
(TEL: +1 (616) 957 6496) (31)

DIXON, Dr. H. B. F., Department of Biochemistry, University of Cambridge, Tennis Court Road, Cambridge CB2 1QW, UK
(TEL: +44 (223) 351781. TELEX: 81240 CAMSPL G, ATTN. H. DIXON, BIOCHEMISTRY) (12, 13)

DOBRESCU, Dr. E. V., Institutul de Cercetări Chimice (ICECHIM), 202 Splaiul Independenţei, sect. 6, R-77208 Bucureşti, Romania
(TEL: 156705. TELEX: 1174 MCHIM R) (23, 25)

DOERFFEL, Prof. K., Sektion Chemie, Technische Hochschule "Carl Schorlemmer" Leuna-Merseburg, Geusaer Strasse, DDR-4200 Merseburg 6, German Democratic Republic
(TELEX: 471320 THLM DD) (28)

DOLEV, Dr. A., Koor Foods Ltd., POB 1597, Haifa 31015, Israel
(TEL: +972 (4) 644328) (33)

DONARUMA, Dr. L. G., Polytechnic Institute of New York, 333 Jay Street, Brooklyn, New York 11201, USA
(TEL: +1 (212) 643 2315) (24)

DONNET, Prof. J. B., Centre de Recherches sur la Physico-Chimie des Surfaces Solides, Centre National de la Recherche Scientifique, 24 avenue du Président Kennedy, F-68200 Mulhouse, France
(TEL: +33 (89) 42 03 55. TELEX: 890032 CNRS-CRO F) (10, 13, 20)

DORNEMANN, Dr. A., Bayer AG, AC-F Untersuchungslaboratorium, Postfach 166, D-4150 Krefeld 11, Federal Republic of Germany
(TEL: +49 (2151) 885375. TELEX: 853401 D) (37)

DOYLE, Prof. M. P., Department of Chemistry, Hope College, Holland, Michigan 49423, USA
(TEL: +1 (616) 392 5111) (21, 22)

DRAŽIĆ, Prof. D., Tehnološko-Metalurški Fakultet, POB 494, Karnejijeva 4, YU-11000 Beograd, Yugoslavia
(TEL: +38 (11) 328320) (16)

DROESCHER, Dr. M., Chemische Werke Hüls AG, Postfach 1320, D-4370 Marl-Hüls 1, Federal Republic of Germany
(TELEX: 829211-0 CW D) (26)

DROST, Dr. H., Zentralinstitut für Physikalische Chemie, Akademie der Wissenschaften der DDR, Rudower Chaussee 5, DDR-1199 Berlin-Adlershof, German Democratic Republic
(TELEX: 112537 ADWAD DD) (15)

DROWART, Prof. J., Laboratorium voor Fysische Chemie, Vrije Universiteit Brussel, Pleinlaan 2, B-1050 Brussel, Belgium
(TEL: +32 (2) 641 34 31. TELEX: 61051 VUBCO B) (20)

DULOG, Prof. L., Forschungsinstitut für Pigmente und Lacke eV, Wiederholdstrasse 10/1, D-7000 Stuttgart 1, Federal Republic of Germany
(TEL: +49 (711) 297549) (24, 25)

DURIG, Prof. J. R., Department of Chemistry, University of South Carolina, Columbia, South Carolina 29208, USA
(TEL: +1 (803) 777 2505. TELEX: 573414 UNIVSCAROL) (17)

DURST, Dr. R. A., Chemistry Building A113, Center for Analytical Chemistry, National Bureau of Standards, US Department of Commerce, Washington, DC 20234, USA
(TEL: +1 (301) 921 2446. TELEX: 898493 GARG) (29)

DUTKA, Dr. F., Magyar Tudományos Akadémia, Központi Kémiai Kutató Intézete, Pusztaszeri utca 59-67, H-1525 Budapest, Hungary
(TEL: +36 (1) 226 686/353 735. TELEX: 226686 KEMIA H) (34)

DUYCKAERTS, Prof. G., Bâtiment 16, Laboratoire de Chimie Analytique et Radiochimie, Université de Liège, B-4000 Sart Tilman-Liège, Belgium
(TEL: +32 (41) 43 17 37. TELEX: 41397 UNIVLG B) (27)

DYBKAER, Dr. R., Department of Clinical Chemistry, Frederiksberg Hospital, Nordre Fasanvej 59, DK-2000 København F, Denmark
(TEL: +45 (1) 34 77 11) (35, 36)

EASTMOND, Dr. G. C., Donnan Laboratories, Department of Inorganic, Physical and Industrial Chemistry, University of Liverpool, POB 147, Grove Street, Liverpool L69 3BX, UK
(TEL: +44 (51) 709 6022. TELEX: 627095 G) (24, 26)

EATHERLY, Dr. W. B., Materials Division, Oak Ridge National Laboratory, POB 10, Oak Ridge, Tennessee 37830, USA
(TELEX: 81572176 USDOE OKRE) (20)

EATON, Dr. D. F., Central Research and Development Department,
E. I. du Pont de Nemours & Co. Inc., Experimental Station, Building 328,
Wilmington, Delaware 19898, USA
(TEL: +1 (302) 772 2181. TELEX: 835420 DUPONT) (22)

EBBIGHAUSEN, Dr. V., Bayer AG, Zentrale Forschung und Entwicklung, Zentrale Analytik,
D-5090 Leverkusen-Bayerwerk, Federal Republic of Germany
(TEL: +49 (214) 305551. TELEX: 85103 0 BY D) (37)

EBEL, Prof. H., Institut für Technische Physik, Technische Universität Wien,
Karlsplatz 13, A-1040 Wien, Austria
(TEL: +43 (222) 56014440) (28)

ECONOMY, Dr. J., IBM Research Laboratory, 5600 Cottle Road, San Jose,
California 95193, USA
(TEL: +1 (408) 256 7988) (23)

EGAN, Dr. H., 5 Highland Road, Amersham, Buckinghamshire HP7 9AP, UK
(TEL: +44 (2403) 28629) (10, 11, 12, 32)

EISENBERG, Prof. H., Department of Polymer Research, Weizmann Institute of Science,
Rehovot 76100, Israel
(TEL: +972 (54) 83252. TELEX: 361900 WIX IL) (13)

ELGERT, Dr., Institut für Physikalische Chemie der Rheinisch-Westfälischen Technischen
Hochschule Aachen, Templergraben 59, D-5100 Aachen, Federal Republic of Germany (25)

EL-SEWEFY, Dr. M. S. E., Egyptian Company for Refractories, 23 Talaat Harb Street,
Cairo, Arab Republic of Egypt (20)

ELYASHÉVICH, Prof. M. A., Faculty of Physics, Byelorussian State University,
Minsk, USSR (17)

ENGLE, Dr. G. B., Graphite Development Branch, General Atomic Co., POB 81608,
San Diego, California 92138, USA
(TEL: +1 (714) 455 3000) (20)

ERIKSSON, Prof. C., SIK-Svenska Livsmedelsinstitutet, Box 5401, S-402 29 Göteborg, Sweden
(TEL: +46 (31) 400120. TELEX: 21651 SIK S) (32)

ERTOLA, Dr. R. J., Facultad de Ciencias Exactas, Centro de Investigacion y Desarrollo
de Fermentaciones Industriales (CINDEFI), Calle 47 y 115, 1900 La Plata, Argentina
(TEL: +54 (21) 33794) (32)

ESSER, Dr. H. O., CIBA-GEIGY Ltd., Agricultural Division, R-1093.3.45,
CH-4002 Basel, Switzerland
(TEL: +41 (61) 37 40 70. TELEX: 62355 CH) (33)

ETTRE, Dr. L. S., Chromatography Division, Perkin-Elmer Corpn., Main Avenue, Norwalk,
Connecticut 06856, USA
(TEL: +1 (203) 762 1000) (28)

EVERETT, Prof. D. H., Department of Physical Chemistry, School of Chemistry,
University of Bristol, Cantock's Close, Bristol BS8 1TS, UK
(TEL: +44 (272) 24161. TELEX: 444174 BUPHYS G) (18)

EWERS, Dr. U., Medizinisches Institut für Umwelthygiene an der Universität Düsseldorf,
Postfach 5634, D-4000 Düsseldorf 1, Federal Republic of Germany
(TEL: +49 (211) 3389372) (37)

EWING, Dr. M. B., Department of Chemistry, University College London,
20 Gordon Street, London WC1H 0AJ, UK
(TEL: +44 (1) 387 7050. TELEX: 28722 G) (16)

EYBL, Dr. V., Department of Pharmacology, Charles University School of Medicine,
Karlovarska 48, CS-301 66 Plzeň, Czechoslovakia (37)

EYSSELTOVA, Dr. JITKA, Faculty of Science, Charles University, Albertov 2030,
CS-128 40 Prague 2, Czechoslovakia (31)

FAGANDINI, Dr. D. A. A., 6 Alleyn Park, Dulwich, London SE21 8AE, UK
(TEL: +44 (1) 670 0547) (10)

FALLAB, Prof. S., Institut für Anorganische Chemie, Universität Basel,
Spitalstrasse 51, CH-4056 Basel, Switzerland
(TEL: +41 (61) 57 15 57) (19)

FARINA, Prof. M., Facoltà di Scienze, Università di Milano, Via Saldini 50,
I-20133 Milano, Italy
(TEL: +39 (2) 2362820) (23)

FARSANG, Dr. Gy., Szervetlen és Analitikai Kémiai Tanszék, Eötvös Loránd
Tudományegyetem, POB 123, H-1443 Budapest, Hungary
(TEL: +36 (1) 189 833/180 820) (29)

FATOU, Prof. J. G., Instituto de Plasticos y Caucho, Consejo Superior
de Investigaciones Científicas, Juan de la Cierva 3, Madrid 6, Spain
(TEL: +34 (1) 2622900. TELEX: 42182 CSIC E) (23)

FAUCHAIS, Prof. P., Laboratoire de Thermodynamique, UER des Sciences,
Université de Limoges, 123 rue Albert Thomas, F-87060 Limoges Cédex, France
(TEL: +33 (55) 77 69 50) (15)

FAULKNER, Prof. L. R., Department of Chemistry, University of Texas at Austin,
Austin, Texas 78712, USA
(TEL: +1 (512) 471 9495. TELEX: 9108741305 UTINTERN AUS) (16)

FEDELI, Prof. E., Stazione Sperimentale per le Industrie degli Oli e dei Grassi,
Via Giuseppe Colombo 79, I-20133 Milano, Italy
(TEL: +39 (2) 2361051) (33)

FEIN, Dr. A. E., Fein-Marquart Associates Inc., 7215 York Road, Baltimore,
Maryland 21212, USA
(TEL: +1 (301) 821 5980) (28)

FELL, Dr. G. S., Department of Biochemistry, Royal Infirmary, Glasgow G4 0SF, UK
(TEL: +44 (41) 552 3535) (37, 38)

FÉRARD, Prof. G., Laboratoire de Biochimie Appliquée, Faculté de Pharmacie,
Université Louis Pasteur, BP 10, 74 route du Rhin, F-67048 Strasbourg Cédex, France
(TEL: +33 (88) 66 90 77/00) (35, 36)

FERNÁNDEZ-PRINI, Dr. R., Departamento Química de Reactores, Comisión Nacional de
Energía Atómica, Avenida del Libertador 8250, 1429 Capital Federal - CP, Argentina
(TEL: +54 (1) 707711. TELEX: 21388 PREAT AR) (30)

FIAT, Prof. D., Medical Center, University of Illinois, POB 6998, Chicago,
Illinois 60680, USA
(TEL: +1 (312) 996 7609) (7)

FINK-JENSEN, Mr. P. H., Athensvej 28, DK-2300 København S, Denmark (25)

FIORELLI, Mr. A., Technology for Coating Resins, Monsanto, 730 Worchester Street,
Indian Orchard, Springfield, Massachusetts 01151, USA (25)

FIRESTONE, Dr. D., HFF 426, Bureau of Foods, Food and Drug Administration,
US Department of Health and Human Services, 200 C Street SW, Washington, DC 20204, USA
(TEL: +1 (202) 245 1381. TELEX: 7108229530) (32)

FISCHBACH, Prof. D. B., Department of Mining, Metals and Ceramic Engineering, FB10,
University of Washington, Seattle, Washington 98195, USA (20)

FISCHER, Prof. E., Department of Structural Chemistry, Weizmann Institute of Science,
Rehovot 76100, Israel
(TELEX: 361900 WIX IL) (22)

FISCHER, Dr. N., Chloe Chimie, BP 35, quai Louis Aulagne, F-69190 Saint Fons, France
(TEL: +33 (7) 867 81 08) (25)

FITZER, Prof. E., Institut für Chemische Technik der Universität Karlsruhe, Kaiserstrasse 12, D-7500 Karlsruhe 1, Federal Republic of Germany
(TEL: +49 (721) 608 2120/2121/691 226. TELEX: 7826521 D) (20)

FLAMM, Dr. D. L., Bell Laboratories, 600 Mountain Avenue, Murray Hill, New Jersey 07974, USA
(TEL: +1 (201) 582 5306. TELEX: 13-8650) (15)

FLEISCHER, Dr. M., Chemische Abteilung, Staatliches Institut für Hygiene und Infektionskrankheiten, Malstatterstrasse 17, D-6600 Saarbrücken, Federal Republic of Germany (37)

FLEISSNER, Dr. M., Hoechst AG, Kunststoff Forschung H, Postfach 800320, D-6230 Frankfurt/Main 80, Federal Republic of Germany
(TEL: +49 (611) 3053422. TELEX: 41234 HOEAG D) (24)

FLUCK, Prof. E., Gmelin-Institut für Anorganische Chemie und Grenzgebiete der Max-Planck-Gesellschaft, Carl Bosch Haus, Postfach 900467, Varrentrappstrasse 40-42, D-6000 Frankfurt/Main 90, Federal Republic of Germany
(TEL: +49 (611) 791 7210. TELEX: 412526 GMELI D) (11, 19)

FLYNN, Dr. J. H., Polymer Standards and Science, National Bureau of Standards, US Department of Commerce, Washington, DC 20234, USA
(TELEX: 898493 GARG) (26)

FODOR-CSÁNYI, Dr. PIROSKA, Magkémiai Laboratórium, Eötvös Loránd Tudományegyetem, Puskin utca 11-13, H-1088 Budapest VIII, Hungary
(TEL: +36 (1) 187 946. TELEX: 225467 ELTE H) (19)

FOGG, Dr. P. G. T., School of Chemistry, Polytechnic of North London, Holloway, London N7 8DB, UK
(TEL: +44 (1) 607 2789. TELEX: 25228 G) (30)

FORSEN, Prof. S., Chemical Center, Physical Chemistry 2, University of Lund, Box 740, S-220 07 Lund 7, Sweden
(TEL: +46 (46) 108245. TELEX: S-33533 LUNIVER) (17)

FOSS, Dr. O. P., Sentrallaboratoriet, Ullevål Sykehus, Oslo 1, Norway (36)

FOX, Dr. R. B., Code 6120, US Naval Research Laboratory, Department of the Navy, Washington, DC 20375, USA
(TEL: +1 (202) 767 2130. TELEX: WU 892778 NRL SUPPLY WSH) (12)

FRANKE, Dr. J. P., Department of Toxicology, Laboratorium voor Farmaceutische en Analytische Chemie, Rijksuniversiteit, Antonius Deusinglaan 2, 9713 AW Gröningen, Netherlands
(TEL: +31 (50) 116892/335-6) (37)

FRANKVOORT, Dr. W., Ministerie van Economische Zaken, Dienst van het Ijkwezen, Postbus 654, 2600 AR Delft, Netherlands
(TEL: +31 (15) 569271) (29)

FRANZINI, Dr. C., Laboratorio Ricerche Cliniche, Ospedale "Filippo del Ponte", Via Filippo del Ponte 19, I-21100 Varese, Italy
(TEL: +39 (332) 285300) (35)

FRASER, Dr. C. G., Department of Biochemical Medicine, Ninewells Hospital and Medical School, Dundee DD1 9SY, UK
(TEL: +44 (382) 60111) (36)

FRAZIER, Prof. J. M., Department of Environmental Health Sciences, John Hopkins University, 415 N. Wolfe Street, Baltimore, Maryland 21205, USA
(TEL: +1 (301) 955 3045) (37)

FREEMAN, Prof. R. D., Department of Chemistry, Oklahoma State University, Stillwater, Oklahoma 74078, USA
(TEL: +1 (405) 624 6676) (15)

FREHSE, Dr. H., Bayer AG, Sparte Pflanzenschutz, Anwendungstechnik CE, Metabolismus und Rückstände, Gebäude W 7, D-5090 Leverkusen-Bayerwerk, Federal Republic of Germany
(TEL: +49 (214) 3072262. TELEX: 8510881 BY D) (10, 32)

FREI, Prof. R. W., Vakgroep Algemene en Analytische Chemie, Subfaculteit der Scheikunde, Vrije Universiteit, De Boelelaan 1083, 1081 HV Amsterdam, Netherlands
(TEL: +31 (20) 5485379. TELEX: 11329 NL) (28)

FREISER, Prof. H., Department of Chemistry, College of Liberal Arts, University of Arizona, Tucson, Arizona 85721, USA
(TEL: +1 (602) 884 3484) (27, 28)

FRITSCHE, Jr., Dr. H. A., University of Texas System Cancer Center, M. D. Anderson Hospital and Tumor Institute, Texas Medical Center, 6723 Bertner Avenue, Houston, Texas 77030, USA
(TEL: +1 (713) 792 6320. TELEX: 910 881-1556) (36)

FRITSCHE, Dr. W., Gesellschaft Deutscher Chemiker, Carl Bosch Haus, Postfach 900440, Varrentrappstrasse 40-42, D-6000 Frankfurt/Main 90, Federal Republic of Germany
(TEL: +49 (611) 791 7320. TELEX: 412526 GMELI D FÜR GDCH) (5, 12)

FRITZ, Prof. H.-G., Institut für Kunststofftechnologie der Universität Stuttgart, Boblinger Strasse 70, D-7000 Stuttgart 1, Federal Republic of Germany
(TEL: +49 (711) 665314) (24)

FROMENT, Prof. G. F., Laboratorium voor Petrochemische Techniek, Rijksuniversiteit te Gent, Krijgslaan 271, B-9000 Gent, Belgium
(TEL: +32 (91) 22 57 15. TELEX: 11793 UB GENT B) (17, 18)

FU, Prof. J., Beijing Institute of Chemical Technology, Heping Street, Beijing, China (30)

FUGER, Prof. J., Bâtiment 16, Laboratoire de Chimie Analytique et Radiochimie, Université de Liège, B-4000 Sart Tilman-Liège, Belgium
(TELEX: 41397 UNIVLG B) (31)

FUKUI, Prof. S., Department of Industrial Chemistry, Faculty of Engineering, Kyoto University, Yoshida-machi, Sakyo-ku, Kyoto, 606 Japan
(TEL: +81 (75) 751 2111. TELEX: 5422693 LIBKYU J) (33)

FUNKE, Prof. W., Forschungsinstitut für Pigmente und Lacke eV, Wiederholdstrasse 10/1, D-7000 Stuttgart 1, Federal Republic of Germany
(TEL: +49 (711) 297549) (25)

FUWA, Prof. K., Department of Chemistry, Faculty of Science, University of Tokyo, 3-1 Hongo 7-chome, Bunkyo-ku, Tokyo, 113 Japan
(TEL: +81 (3) 812 2111. TELEX: 25510 UNITOKYO J) (27)

GAD, Prof. A. M., Fats and Oils Research Laboratory, National Research Centre, Dokki, Cairo, Arab Republic of Egypt (33)

GALAS, Prof. E., Institute of Technical Biochemistry, Łódź Technical University, ul. Gdańska 162/168, PL-90 924 Łódź, Poland (33)

GALLOT, Mme., Centre de Recherches sur les Macromolécules du Centre National de la Recherche Scientifique, 6 rue Boussingault, F-67083 Strasbourg Cédex, France
(TEL: +33 (88) 61 19 19) (25)

GALUS, Prof. Z., Wydział Chemii, Uniwersytet Warszawski, ul. Pasteura 1, PL-02 093 Warszawa, Poland
(TEL: +48 (22) 220211. TELEX: 815439 UW PL) (29)

GAMBLE, Dr. D. S., Environmental Chemistry Section, Chemistry and Biology Research Institute, Agriculture Canada, Ottawa, Ontario K1A 0C6, Canada
(TELEX: 053 3283 CANAGRI CA) (29)

GARAJ, Prof. J., Katedra Analytickej Chémie, Chemiskotechnologická Fakulta, Slovenská Vysoká Škola Technická, Jánská 1, CS-812 37 Bratislava, Czechoslovakia
(TEL: +42 (427) 56043. TELEX: 93521 C) (27)

GARDNER, Prof. MARJORIE, Department of Chemistry, University of Maryland, College Park, Maryland 20742, USA
(TEL: +1 (301) 454 2310) (11)

GARN, Prof. P. D., Department of Chemistry, University of Akron, Akron, Ohio 44325, USA (26)

GASSIOT MATAS, Prof. M., Instituto Quimico de Sarrià, Barcelona 17, Spain (33)

GAT, Prof. J. R., Department of Isotope Research, Weizmann Institute of Science, Rehovot 76100, Israel
(TELEX: 361900 WIX IL) (19)

GAUR, Prof. H. C., Department of Chemistry, Delhi University, Delhi-110007, India (29)

GEARY, Dr. T. D., Division of Clinical Chemistry, IMVS, Frome Road, Adelaide, Southern Australia 5000, Australia
(TEL: +61 (8) 2287911. TELEX: AA82647 MEDVET) (35)

GEIL, Dr. P. H., Polymer Group, University of Illinois at Urbana-Champaign, 1304 W Green Street, Urbana, Illinois 61801, USA
(TEL: +1 (217) 333 1441) (24)

GEORGE, Prof. M. V., Department of Chemistry, Indian Institute of Technology, Kanpur-208016, India (22)

GERMAIN, Dr. R., Medical Department, Société Métallurgique le Nickel, PO Box 5, Noumea, New Caledonia (37)

GERRANS, Prof. G. C., Department of Chemistry, University of the Witwatersrand, 1 Jan Smuts Avenue, Johannesburg 2001, Republic of South Africa
(TEL: +27 (11) 7162183. TELEX: 4-22460 SA) (12)

GERRARD, Dr. W., School of Chemistry, Polytechnic of North London, Holloway, London N7 8DB, UK
(TEL: +44 (1) 607 2789. TELEX: 25228 G) (30)

GERSTER, Dr. A. C., Kunstharz-Abteilung, Dr. Walter Mäder AG, CH-8956 Killwangen, Switzerland (25)

GETZEN, Prof. F. W., Department of Chemistry, North Carolina State University, Raleigh, North Carolina 27650, USA
(TEL: +1 (919) 737 3154) (30)

GEUSKENS, Prof. G., Faculté des Sciences, Université Libre de Bruxelles, Campus Plaine 206/1, Boulevard du Triomphe, B-1050 Bruxelles, Belgium
(TEL: +32 (2) 640 00 15. TELEX: 23069 UNILIBRE BRUX B) (11)

GEVANTMAN, Dr. L. H., Office of Standard Reference Data, A318 Physics Building, National Bureau of Standards, US Department of Commerce, Washington, DC 20234, USA
(TEL: +1 (301) 921 3442. TELEX: 898493 GARG) (30)

GHANEM, Prof. N. A., c/o Prof. A. B. Fayez, Academy of Scientific Research and Technology, 101 Kasr El-Ainy Street, Cairo, Arab Republic of Egypt (24)

GHAZALI BIN HJ. ABDUL RAHMAN, Prof. M., PUSPATI, No. 4, Jalan 1/4, Bandar Baru Bangi, Selangor, Malaysia
(TEL: +60 (3) 341644) (19)

GHIJSELS, Mr. A., Koninklijke/Shell Laboratorium, Shell Research BV, Postbus 3003, Amsterdam-N, Netherlands
(TEL: +31 (20) 303145. TELEX: 11224 KSLA NL) (24)

GHOSE, Prof. T. K., Biochemical Engineering Research Centre, Indian Institute of Technology, Hauz Khas, New Delhi-110016, India
(TEL: +91 (11) 651880. TELEX: 3687 ND IN) (32)

GIESBRECHT, Prof. E., Instituto de Química, Universidade de São Paulo, Caixa Postal 20780, 01000 São Paulo - SP, Brazil
(TEL: +55 (11) 210 2122. TELEX: 21519 BR) (11)

GIJBELS, Prof. R., Trace Analysis Centre, Departement Scheikunde, Universitaire Instelling Antwerpen, Universiteitsplein 1, B-2610 Wilrijk, Belgium
(TEL: +32 (31) 28 25 28. TELEX: 33646 B) (27)

GILBERT, Dr. B., Bâtiment B 16, Laboratoire de Chimie Analytique et Radiochimie, Université de Liège, B-4000 Sart Tilman-Liège, Belgium
(TELEX: 41397 UNIVLG B) (29)

GILEADI, Prof. E., Department of Chemistry, Tel-Aviv University, Tel-Aviv, Israel
(TEL: +972 (3) 420268/411288. TELEX: 69978 VERSY IL) (16)

GILES, Jr., Dr. P. M., Chemical Abstracts Service, POB 3012, 2540 Olentangy River Road, Columbus, Ohio 43210, USA
(TEL: +1 (614) 421 3632. TELEX: 684-2086) (21)

GILL, Prof. S. J., Department of Chemistry, University of Colorado, Boulder, Colorado 80302, USA
(TEL: +1 (303) 492 8273. TELEX: 910 940 3441) (13)

GILLES, Prof. P. W., Department of Chemistry, University of Kansas, Lawrence, Kansas 66045, USA
(TEL: +1 (913) 864 3829. TELEX: 910 749 6572) (20)

GIOLITO, Prof. I., Instituto de Química, Universidade de São Paulo, Caixa Postal 20780, 01000 São Paulo - SP, Brazil
(TELEX: 21519 BR) (28)

GIRARD, Mr. G., Bureau International des Poids et Mesures, pavillon de Breteuil, 12 bis Grande Rue, F-92310 Sèvres, France
(TEL: +33 (1) 534 00 51) (15)

GITELMAN, Dr. H. J., Department of Medicine, Division of Nephrology, University of North Carolina, Old Clinic Building 226H, Chapel Hill, North Carolina 27514, USA
(TEL: +1 (919) 966 2561) (38)

GLASER, Mr. M. A., Midland Division, Dexter Corpn., 1-7 East Water Street, Waukegan, Illinois 60085, USA (25)

GLATTES, Dipl. Ing. H., ICC General Sekretariat, Schmidgasse 3-7, A-2320 Schwechat, Austria
(TEL: 777202. TELEX: 3316 A) (5)

GOCHMAN, Dr. N., Clinical Systems Division, Beckman Instruments Inc., 200 South Kraemer Boulevard, Brea, California 92621, USA
(TEL: +1 (714) 993 5321. TELEX: 678413) (35)

GODINHO, Prof. O. E. S., UNICAMP - Caixa Postal 1170, Cidade Universitária Barão Geraldo, 13100 Campinas - SP, Brazil (29)

GODLY, Mr. E. W., Laboratory of the Government Chemist, Cornwall House, Stamford Street, London SE1 9NQ, UK
(TEL: +44 (1) 928 7900. TELEX: 881314 DIHQG LAB GOV CHEM G) (19, 21)

GOEL, Prof. P. S., Department of Chemistry, Indian Institute of Technology, Kanpur-208016, India (19)

GOETHALS, Prof. E., Laboratorium voor Makromoleculaire Scheikunde, Universiteit te Gent, Krijgslaan 271, B-9000 Gent, Belgium
(TEL: +32 (91) 22 57 15. TELEX: 12754 RUGENT B) (23)

GOH, Dr. L. Y., Jabatan Kimia, Universiti Malaya, Lembah Pantai, Kuala Lumpur 22-11, Malaysia
(TEL: +60 (3) 555466. TELEX: 37453 MA) (19)

GOLD, Prof. V., Department of Chemistry, King's College, Strand, London WC2R 2LS, UK
(TEL: +44 (1) 836 5454) (12, 21, 22)

GOLDANSKII, Prof. V. I., Institute of Chemical Physics, Academy of Sciences of USSR, Kosygin Street 4, SU-117977, SZD-1, Moscow, V-334, USSR
(TEL: +7 (095) 139 7300) (19)

GOLDBACH, Dr. G., Chemische Werke Hüls AG, ZB Forschung und Entwicklung, Postfach 1320, D-4370 Marl-Hüls 1, Federal Republic of Germany
(TEL: +49 (2365) 495210. TELEX: 829211-0 CW D) (24)

GOLDBERG, Prof. E. D., Geological Research Division A-020, Scripps Institution of Oceanography, La Jolla, California 92093, USA
(TEL: +1 (619) 452 2407) (29)

GOLDSTEIN, Prof. L., Department of Biochemistry, Tel-Aviv University, Tel-Aviv, Israel
(TELEX: 69978 VERSY IL) (32)

GOLOVLEVA, Dr. L. A., Laboratory of Enzymatic Degradation of Organic Compounds, Institute of Biochemistry and Physiology of Microorganisms, Academy of Sciences of USSR, SU-142292 Pushchino, Moscow Region, USSR
(TEL: +7 (095) 295 7448) (33)

GOMIŠČEK, Prof. S., Oddelek za Kemijo, Fakultete za Naravoslovje in Tehnologijo, Univeze v Ljubljana, POB 537, Murnikova 6, YU-61001 Ljubljana, Yugoslavia (37)

GOOD, Dr. MARY L., UOP Corporate Research Center, UOP Inc., Ten UOP Plaza, Des Plaines, Illinois 60016, USA
(TEL: +1 (312) 391 3331. TELEX: 253 285 UOP-D) (10, 19)

GORBACH, Dr. S., Hoechst AG, Analytisches Laboratorium, Postfach 800320, D-6230 Frankfurt/Main 80, Federal Republic of Germany
(TEL: +49 (611) 3056012. TELEX: 41234 HOEAG D) (32, 33)

GORNALL, Dr. A., Department of Clinical Biochemistry, Banting Institute, University of Toronto, Toronto, Ontario M5G 1L5, Canada (36)

GRACIAN TOUS, Prof. J., Instituto de la Grasa y sus Derivados, Consejo Superior de Investigaciones Científicas, Avenida Padre García Tejero 4, Apartado 1078, Sevilla 12, Spain
(TEL: +34 (54) 611550) (33)

GRANDJEAN, Dr. Ph., Institut for Samfundsmedicin, Department of Environmental Medicine, Odense Universitet, J. B. Winsløws Vej 19, DK-5000 Odense C, Denmark
(TEL: +45 (9) 11 33 33) (37)

GRASSELLI, Dr. JEANNETTE G., Standard Oil Co. (Ohio), 4440 Warrensville Center Road, Cleveland, Ohio 44128, USA
(TEL: +1 (216) 581 5825. TELEX: 6873120) (17)

GRASSERBAUER, Prof. M., Institut für Analytische Chemie, Technische Universität Wien, Getreidemarkt 9/151, A-1060 Wien, Austria
(TEL: +43 (222) 56014847/4837) (12, 27, 28)

GRAVELLE, Dr. P. C., Institut de Recherches sur la Catalyse du Centre National de la Recherche Scientifique, 2 avenue Albert Einstein, F-69626 Villeurbanne Cédex, France
(TEL: +33 (7) 893 34 71. TELEX: 900225 CNRS VILRB F) (17, 18)

GRAY, Dr. P. P., School of Biological Technology, University of New South Wales, PO Box 1, Kensington, New South Wales 2033, Australia
(TEL: +61 (2) 663 0351. TELEX: AA26054) (32)

GREENHALGH, Dr. R., Chemistry and Biology Research Institute, Agriculture Canada, Ottawa, Ontario K1A 0C6, Canada
(TEL: +1 (613) 995 3700. TELEX: 053 3283 CANAGRI CA) (32, 33)

GREENWOOD, Prof. N. N., Department of Inorganic and Structural Chemistry, University of Leeds, Leeds LS2 9JT, UK
(TEL: +44 (532) 431751) (19)

GRENTHE, Prof. I., Department of Inorganic Chemistry, Royal Institute of Technology, S-100 44 Stockholm 70, Sweden
(TEL: +46 (8) 7878144. TELEX: 10389 KTHB STOCKHOLM S) (29, 30)

GRIBOV, Prof. L. A., V I Vernadskii Institute of Geochemistry and Analytical Chemistry, Academy of Sciences of USSR, Kosygin Street 19, SU-117975, SZD-1, Moscow, V-334, USSR
(TEL: +7 (095) 137 6371) (17)

GRIEPINK, Dr. B., Zuider-Kerkstraat 4, 3421 AL Oudewater, Netherlands
(TEL: +31 (3486) 2460) (27, 28)

GRIES, Dr. W., National Institute for Materials Research, Council for Scientific and Industrial Research, PO Box 395, Pretoria 0001, Republic of South Africa (TELEX: 3-630 SA) (27)

GRIFFIN, Dr. R. M., ESA Laboratories Inc., 43 Wiggins Avenue, Bedford, Massachusetts 01730, USA (TEL: +1 (617) 275 0100. TELEX: 923344) (37, 38)

GRIFFITHS, Prof. P. D., Department of Biochemical Medicine, Ninewells Hospital and Medical School, Dundee DD1 9SY, UK (TEL: +44 (382) 60111) (35, 36)

GRIMMER, Prof. G., Bebelallee 30A, D-2000 Hamburg 60, Federal Republic of Germany (TEL: +49 (4102) 62155) (32)

GRITZNER, Prof. G., Institut für Chemische Technologie Anorganischer Stoffe, Johannes Kepler Universität Linz, A-4040 Linz-Auhof, Austria (TEL: +43 (732) 231381) (16)

GRONOWITZ, Prof. S., Chemical Center, Organic Chemistry 1, University of Lund, Box 740, S-220 07 Lund 7, Sweden (TEL: +46 (46) 108125. TELEX: S-33533 LUNIVER) (21)

GROSS, Prof. M., Laboratoire d'Electrochimie et Chimie Physique du Corps Solide, ERA au CNRS n° 468, Université Louis Pasteur, 4 rue Blaise Pascal, F-67000 Strasbourg Cédex, France (TEL: +33 (88) 61 48 30. TELEX: 890032 CNRS-CRO F) (29)

GROVES, Dr. D. J., Petrochemicals and Plastics Division, Imperial Chemical Industries PLC, Wilton Centre, POB 90, Wilton, Middlesborough, Cleveland TS6 8JE, UK (TEL: +44 (642) 455522. TELEX: 587461 G) (24)

GRUSHKA, Prof. E., Department of Inorganic and Analytical Chemistry, Hebrew University of Jerusalem, Jerusalem 91904, Israel (TEL: +972 (2) 585303. TELEX: 25391 HU IL) (28, 29)

GRÜTZMACHER, Prof. H.-F., Fakultät für Chemie, Universität Bielefeld, Postfach 8640, Universitätsstrasse 25, D-4800 Bielefeld, Federal Republic of Germany (TEL: +49 (521) 1062072) (17)

GUERRERO, Prof. A. H., Avenida Santa Fe 2879, 1° B, 1425 Buenos Aires, Argentina (TEL: +54 (1) 8215249) (11)

GUILBAULT, Prof. G. G., Department of Chemistry, University of New Orleans, Lake Front, New Orleans, Louisiana 70122, USA (TEL: +1 (504) 283 0323) (27, 28)

GUMINSKI, Dr. C., Wydział Chemii, Uniwersytet Warszawski, ul. Pasteura 1, PL-02 093 Warszawa, Poland (TELEX: 815439 UW PL) (31)

GURVICH, Prof. L. V., Institute for High Temperatures, Academy of Sciences of USSR, Korovinskoye Road, SU-127412 Moscow, USSR (TEL: +7 (095) 485 9572) (20)

GUTHRIE, Prof. R. D., Department of Chemistry, College of Arts and Sciences, University of Kentucky, Lexington, Kentucky 40506, USA (TEL: +1 (606) 258 2200) (21)

GUTMANN, Dr. H., Pharma Forschung C, F. Hoffmann-La Roche & Co. AG, CH-4002 Basel, Switzerland (TEL: +41 (61) 27 24 63. TELEX: 62292/62072 CH) (21)

GUYOT, Dr. A., Laboratoire de Matériaux Organiques du Centre National de la Recherche Scientifique, BP 24, F-69390 Vernaison, France (TEL: +33 (7) 802 10 19. TELEX: 900225 CNRS VILRB F) (25)

HABER, Prof. J., Instytut Katalizy i Fizykochemii Powierzchni, Polska Akademia Nauk, ul. Niezapominajek, PL-30 239 Kraków, Poland (TEL: +48 (94) 225727. TELEX: 0325702 IKAT PL) (13, 17, 18)

HABRAKEN, Dr. CLARISSE, c/o Koninklijke Nederlandse Chemische Vereniging, Burnierstraat 1, 2596 HV Den Haag, Netherlands (12)

HADJIVASSILIOU, Dr. A. G., 66 Spyrou Mercouri Street, Athens 116 34, Greece (TEL: +30 (1) 4518411/7232921) (36)

HADTZCHRISTIDIS, Mrs. C., Environmental Pollution Control Project, 147 Patission, Athens, Greece
(TEL: +30 (1) 8650053/334. TELEX: 216028 DYPP GR) (37)

HAECKEL, Prof. R., Institut für Laboratoriumsmedizin, Zentralkrankenhaus, St. Jürgenstrasse, D-2800 Bremen, Federal Republic of Germany
(TEL: +49 (421) 4973560) (35)

HAGEMANN, Dr. R., Service d'Etudes Analytiques, Centre d'Etudes Nucléaires de Fontenay-aux-Roses, BP 6, F-92260 Fontenay-aux-Roses, France
(TEL: +33 (1) 654 81 30/81 31. TELEX: 270898 ENERGAT FNAYR F) (19, 28)

HAGUENOER, Prof. J. M., Université du Droit et de la Santé de Lille, UER de Médicine du Travail, BP 259, place de Verdun, F-59045 Lille Cédex, France (37)

HAKKI, Prof. W., c/o Dr. M. Adnan Rifai, Secretary General, Supreme Council of Sciences, POB 4762, Damascus, Syrian Arab Republic (12)

HÁLA, Dr. J., Fakulta Anorganické Chemie, Universita J. E. Purkyně, CS-611 37 Brno, Czechoslovakia (30)

HALEVI, Prof. E. A., Department of Chemistry, Technion-Israel Institute of Technology, Technion City, Haifa 32000, Israel
(TEL: +972 (4) 293745. TELEX: 46650 TECLI IL) (21)

HAMANN, Prof. K., August-Lämmle-Weg 46, D-7250 Leonberg, Federal Republic of Germany (25)

HAMFELT, Dr. A., Klinisk-Kemiska Laboratoriet, Sundsvalls Sjukhus, S-851 86 Sundsvall, Sweden
(TEL: +46 (60) 181360) (35)

HANLEY, Dr. H. J. M., Thermophysical Properties Division, National Engineering Laboratory, National Bureau of Standards, Boulder, Colorado 80302, USA
(TEL: +1 (303) 497 3320. TELEX: 45 897 SOLTERWARN BDR) (16)

HANSEN, Dr. Ch. M., Scandinavian Paint and Printing Ink Research Laboratory, Agern Alle 3, DK-2970 Hørsholm, Denmark (25)

HANSEN, Dr. K., Danish National Institute of Occupational Health, Baunegaardsvej 73, DK-2900 Hellerup (Copenhagen), Denmark
(TELEX: 16149 AMI DK) (37)

HANSEN, Prof. R. S., Ames Laboratory, Energy and Mineral Resources Research Institute, Iowa State University, Ames, Iowa 50011, USA
(TEL: +1 (515) 294 2770. TELEX: 910 520 1167) (18)

HAO, Dr. P. L. C., Union Chemical Laboratories, Industrial Technology Research Institute, 5th Floor, No. 1, Jen-Ai Road, Sec. 2, Taipei, Taiwan
(TEL: 02 3921250-3) (10)

HARADA, Dr. A., Environmental Pollution Control Division, Sanyo Electric Co. Ltd., 2-18 Keihan-hondori, Moriguchi City, Osaka, Japan
(TEL: +81 (6) 371 4121) (37)

HARDY, Prof. G., Müanyagipari Kutató Intézet, Hungária krt. 114, H-1950 Budapest, Hungary
(TEL: +36 (1) 636 245. TELEX: 226493 MUEKI H) (23)

HARDY, Dr. R. W. F., Central Research and Development Station, E. I. du Pont de Nemours & Co. Inc., Experimental Station, Wilmington, Delaware 19898, USA
(TEL: +1 (302) 772 2181. TELEX: 835420 DUPONT) (13)

HARRIS, Dr. C. K., Department of Laboratories, St. Paul's Hospital, Vancouver, British Columbia V6Z 1Y6, Canada
(TEL: +1 (604) 682 5009) (35)

HARRIS, Prof. R. K., School of Chemical Sciences, University of East Anglia, University Plain, Norwich NR4 7TJ, UK
(TEL: +44 (603) 56161. TELEX: 97154 G) (17)

HARRIS, Prof. W. E., Department of Chemistry, University of Alberta, Edmonton, Alberta T6G 2E1, Canada
(TEL: +1 (403) 432 3254) (28)

HARTMANN, Dr. A., Rothschild Bank AG, Zollikerstrasse 181, CH-8008 Zürich, Switzerland
(TEL: +41 (1) 384 71 11. TELEX: 558511 CH) (12)

HASHIMOTO, Dr. T., c/o The Japan Oil Chemists' Society, 7th Floor, Yushi Kogyo Kaikan, 13-11, Nihonbashi 3-chome, Chuo-ku, Tokyo, 103 Japan
(TEL: +81 (3) 271 7463) (33)

HASTIE, Dr. J., Materials 223/A329, National Bureau of Standards, US Department of Commerce, Washington, DC 20234, USA
(TEL: +1 (301) 921 2859. TELEX: 898493 GARG) (20)

HAUL, Prof. R. A. W., Institut für Physikalische Chemie und Elektrochemie der Technischen Universität Hannover, Callinstrasse 3-3A, D-3000 Hannover 1, Federal Republic of Germany
(TEL: +49 (511) 7622250. TELEX: 923868 TU HAN D) (18)

HAULAIT-PIRSON, Dr. MARIE-CLAIRE, Departement Scheikunde, Laboratorium voor Fysicochemie en Stralingschemie, Katholieke Universiteit te Leuven, Celestijnenlaan 200 F, B-3030 Heverlee, Belgium
(TEL: +32 (16) 20 06 56. TELEX: 23674 KULEUV B) (30)

HAUTFENNE, Dr. A., 22 Avenue des Grenadiers, B-1330 Rixensart, Belgium
(TEL: +32 (10) 41 81 81) (33)

HAVIARAS, Dr. V., 4 Alkmanos Street, Athens 115 28, Greece (35)

HAYAKAWA, Dr. S., Mitsui Petrochemical Industries Ltd., 3-2-5 Kasumigaseki, Chiyoda-ku, Tokyo, 100 Japan
(TEL: +81 (3) 580 3611. TELEX: 22984 MIPECA J) (24)

HAYASHI, Dr. Ch., Department of Laboratory Medicine, Medical School, Osaka University, 1-1-50 Fukushima, Fukushima-ku, Osaka, 553 Japan
(TEL: +81 (6) 451 0051) (35)

HAYDUK, Prof. W., Department of Chemical Engineering, University of Ottawa, Ottawa, Ontario K1N 9B4, Canada
(TEL: +1 (613) 231 4095. TELEX: 053 3333 CA) (30)

HEAD, Dr. A. J., Division of Quantum Metrology, National Physical Laboratory, Department of Industry, Teddington, Middlesex TW11 0LW, UK
(TEL: +44 (1) 977 3222. TELEX: 262344 G) (15)

HEFTER, Dr. G. T., School of Mathematical and Physical Sciences, Murdoch University, Murdoch, Western Australia 6157, Australia
(TELEX: AA92711 MULIB) (29, 30)

HEIMDAHL, Siv.ing. S., A/S Sotungruppen, POB 233, N-3201 Sandefjord, Norway (25)

HEINRICH, Dr. K., Office of Foreign Relations, National Bureau of Standards, US Department of Commerce, Washington, DC 20234, USA
(TELEX: 898493 GARG) (27, 28)

HEITZ, Prof. W., Fachbereich Physikalische Chemie, Polymere, Universität Lahnberge, Gebäude H, Hans-Meerwein-Strasse, D-3550 Marburg/Lahn 1, Federal Republic of Germany
(TEL: +49 (6421) 285776. TELEX: 482372 UMR D) (23)

HELLER, Dr. A., Bell Laboratories, 600 Mountain Avenue, Murray Hill, New Jersey 07974, USA
(TEL: +1 (201) 582 3000. TELEX: 13-8650) (22)

HELLWINKEL, Prof. D., Organisch-Chemisches Institut der Universität Heidelberg, Im Neuenheimer Feld 270, D-6900 Heidelberg 1, Federal Republic of Germany
(TEL: +49 (6221) 562420) (21)

HEMINGWAY, Dr. R. J., Plant Protection Division, Imperial Chemical Industries PLC, Jealott's Hill Research Station, Bracknell, Berkshire RG12 6EY, UK
(TEL: +44 (344) 424701. TELEX: 847556 ICI G) (33)

HENDRIKSE, Mr. P. W., Unilever Research Laboratorium, Postbus 114, 3130 AC Vlaardingen, Netherlands
(TEL: +31 (10) 353000. TELEX: 23261 NL) (33)

HENG, Dr. C. K., MARDI, Peti Surat 2301, Pejabat Besar Pos, Kuala Lumpur, Malaysia
(TEL: +60 (3) 356601. TELEX: 37115 MARDI MA) (34)

HENZE, Dr. G., Bayer AG, IN AP, Gebäude E 41, D-5090 Leverkusen-Bayerwerk, Federal Republic of Germany
(TELEX: 85103 0 BY D) (26)

HEPLER, Prof. L. G., Department of Chemistry, University of Lethbridge, 4401 University Drive, Lethbridge, Alberta T1K 3M4, Canada
(TEL: +1 (403) 329 2304. TELEX: 03-849357 CA) (13)

HERB, Dr. G. K., Bell Laboratories, 555 Union Boulevard, Allentown, Pennsylvania 18013, USA
(TEL: +1 (215) 439 7844) (15)

HERBER, Dr. R. F. M., Coronel Laboratorium, Universiteit van Amsterdam, Eerste Constantijn Huygenstraat 20, Amsterdam, Netherlands
(TEL: +31 (20) 5784072) (37)

HERZBERG, Dr. G., Herzberg Institute of Astrophysics, National Research Council of Canada, Ottawa, Ontario K1A 0R6, Canada
(TEL: +1 (613) 992 2350. TELEX: 053-3715 NRC OTT CA) (17)

HEUMANN, Prof. K. G., Institut für Anorganische Chemie, Universität Regensburg, Universitätsstrasse 31, D-8400 Regensburg, Federal Republic of Germany
(TEL: +49 (941) 943 4448. TELEX: 065658 UNIRE D) (19)

HEUSLER, Prof. K. E., Abteilung Korrosion der Technischen Universität Clausthal, Agricolastrasse 2, D-3392 Clausthal-Zellerfeld, Federal Republic of Germany
(TEL: +49 (5323) 722438. TELEX: 953828 TU CLZ D) (16)

VAN DEN HEUVEL, Dr. C. J. M., AKZO Research Laboratories, Department CRL, PO Box 60, 6800 AB Arnhem, Netherlands
(TEL: +31 (85) 662257. TELEX: 45204 NL) (25)

HICKS, Dr. JOCELYN, Department of Clinical Laboratories, Childrens Hospital, National Medical Center, 111 Michigan Avenue NW, Washington, DC 20010, USA
(TEL: +1 (202) 745 5339) (36)

HIEFTJE, Prof. G. M., Department of Chemistry, Chemistry Building, Indiana University, Bloomington, Indiana 47405, USA
(TEL: +1 (812) 337 2189. TELEX: 272279) (28)

HIGHTOWER, Prof. J. W., Department of Chemical Engineering, Rice University, POB 1892, Houston, Texas 77251, USA
(TEL: +1 (713) 527 8101) (17, 18)

HILDEBRAND, Dr. H. F., Institut de Recherches sur le Cancer de Lille, U.124 INSERM, BP 311, F-59020 Lille Cédex, France
(TEL: +33 (20) 95 43 59) (37)

HILL, Dr. F., Eastern Health Board, City Laboratory, 10 Corn Market, Dublin 8, Ireland
(TEL: +0044 (0001) 116450) (37)

HILL, Dr. J. G., Department of Clinical Chemistry, Hospital for Sick Children, 555 University Avenue, Toronto, Ontario M5G 1X8, Canada (6, 35)

HINRICHSEN, Prof. G., Institut für Nichtmetallische Werkstoffe, Technische Universität Berlin, Englische Strasse 20, D-1000 Berlin 12, Federal Republic of Germany
(TEL: +49 (30) 3144464. TELEX: 184262 TUBLN D) (24, 26)

HINZ, Dr. H.-J., Institut für Biophysik und Physikalische Biochemie, Universität Regensburg, Universitätsstrasse 31, D-8400 Regensburg, Federal Republic of Germany
(TEL: +49 (941) 943 3003. TELEX: 065658 UNIRE D) (13)

HIRAYAMA, Dr. C., Chemical Physics, Research and Development Center, Westinghouse Electric Corpn., 1310 Beulah Road, Pittsburgh, Pennsylvania 15235, USA (31)

HIRAYAMA, Dr. K., Science Information Institute, 720 Soshi, Matsuda, Kanagawa-ken, 258 Japan
(TEL: +81 (465) 82 0678) (21)

HITCHINGS, Mr. T. R., Riccarton High School, Curlett's Road, Christchurch 4, New Zealand
(TEL: +64 (3) 45073) (12)

HJELM, Prof. M., Department of Clinical Biochemistry, Hospital for Sick Children, Great Ormond Street, London WC1N 3JH, UK
(TEL: +44 (1) 405 9200) (14, 35)

HLAVÁČ, Dr. J., Katedra Technologie Silikátů, Vysoká Škola Chemicko-Technologická v Praze, Suchbátarova 5, CS-166 28 Praha 6-Dejvice, Czechoslovakia
(TEL: +42 (2) 320259. TELEX: 122744 VSCH C) (19)

HNOOSH, Dr. M. H., Union of Arab Chemists, POB 8011, Damascus Street, Salihiyah, Baghdad, Iraq
(TEL: 30313) (7)

HOCKING, Dr. M. G., Department of Metallurgy and Materials Science, Imperial College of Science and Technology, Prince Consort Road, South Kensington, London SW7 2BY, UK
(TEL: +44 (1) 589 5111. TELEX: 261503 G) (20)

HOFACKER, Dr. URSULA, Gesellschaft Deutscher Chemiker, Carl Bosch Haus, Postfach 900440, Varrentrappstrasse 40-42, D-6000 Frankfurt/Main 90, Federal Republic of Germany
(TEL: +49 (611) 791 7326. TELEX: 412526 GMELI D FÜR GDCH) (11)

HØGETVEIT, Dr. A., Falconbridge Nikkelverk Aktieselskap, POB 457, N-4601 Kristiansand S, Norway
(TEL: +47 (42) 25500) (37)

HÖGFELDT, Dr. E., Department of Inorganic Chemistry, Royal Institute of Technology, S-100 44 Stockholm 70, Sweden
(TELEX: 10389 KTHB STOCKHOLM S) (29, 30)

HØILAND, Prof. H., Department of Chemistry, University of Bergen, Allégate 41, N-5014 Bergen-Universitet, Norway
(TEL: +47 (5) 320040/3371. TELEX: 42877 N) (30)

HOLDEN, Dr. N. E., Department of Nuclear Energy-197 D, National Nuclear Data Center, Brookhaven National Laboratory, Upton, New York 11973, USA
(TEL: +1 (516) 345 2901. TELEX: 967703 BROOKLAB UPTONY) (19)

HOLLINGWORTH, Dr. R., Department of Entomology, Purdue University, West Lafayette, Indiana 47907, USA
(TEL: +1 (317) 494 4594. TELEX: 272396) (33)

HOLLÓ, Prof. J., Magyar Tudományos Akadémia, Központi Kémiai Kutató Intézete, Pusztaszeri utca 59-67, H-1525 Budapest, Hungary
(TEL: +36 (1) 151 602/352 314. TELEX: 226686 KEMIA H) (32)

HØLMER, Dr. GUNHILD, Department of Biochemistry and Nutrition, Danmarks Tekniske Højskole, Bygning 224, DK-2800 Lyngby, Denmark
(TEL: +45 (2) 88 40 66. TELEX: 37529 DTHDIA DK) (33)

HOLMSTRÖM, Dr. A., Statens Provningsanstalt, Box 857, S-501 15 Borås, Sweden (25)

HOLTRUP, Mr. W., Chemische Werke Hüls AG, ZB Forschung und Entwicklung/Physikalische Chemie, Postfach 1320, D-4370 Marl-Hüls 1, Federal Republic of Germany
(TELEX: 829211-0 CW D) (25)

HOMANN, Prof. K. H., Institut für Physikalische Chemie der Technischen Hochschule Darmstadt, Petersenstrasse 20, D-6100 Darmstadt, Federal Republic of Germany
(TEL: +49 (6151) 16 21 72. TELEX: 419579 D) (15)

HORÁK, Prof. J., Katedra Organichké Technologie, Vysoká Škola Chemicko-Technologická v Praze, Suchbátarova 5, CS-166 28 Praha 6-Dejvice, Czechoslovakia
(TEL: +42 (2) 321551. TELEX: 122744 VSCH C) (11)

HORÁNYI, Dr. G., Magyar Tudományos Akadémia, Központi Kémiai Kutató Intézete, Pusztaszeri utca 59-67, H-1525 Budapest, Hungary
(TEL: +36 (1) 335 350. TELEX: 226686 KEMIA H) (16)

HORIO, Prof. T., Institute for Protein Research, Osaka University, Yamada-oka 1-1, Suita-city, Osaka-fu, 565 Japan
(TEL: +81 (6) 877 5111) (36)

HORVATH, Dr. A. L., ICI Mond Division, POB 8, The Heath, Runcorn WA7 4QD, UK (30)

HORVÁTH, Dr. ZSUZSA, Szervetlen és Analitikai Kémiai Tanszék, Eötvös Loránd Tudományegyetem, POB 123, H-1443 Budapest, Hungary (27)

HORWITZ, Dr. W., HFF 7, Bureau of Foods, Food and Drug Administration, US Department of Health and Human Services, 200 C Street SW, Washington, DC 20204, USA
(TEL: +1 (202) 245 1057. TELEX: 7108229530) (28)

HOUGEN, Dr. J. T., Molecular Spectroscopy Division, National Bureau of Standards, US Department of Commerce, Washington, DC 20234, USA
(TELEX: 898493 GARG) (17)

HOUK, Prof. K. N., Department of Chemistry, University of Pittsburgh, Pittsburgh, Pennsylvania 15260, USA
(TEL: +1 (412) 624 5130. TELEX: 812 466) (22)

HOXTER, Prof. G., Faculdade de Ciências Farmacêuticas, Universidade de São Paulo, Caixa Postal 30786, 05508 São Paulo, Brazil
(TEL: +55 (11) 210 2122) (35)

HSIA, Dr. R. C. H., Syarikat Poly Trades, 8229, Jalan 222, Petaling Jaya, Selangor, Malaysia (27)

HUA, Prof. T., Department of Chemistry, Peking University, Beijing, China
(TEL: 282471. TELEX: 222 39 PKUNI CN) (11)

HUANG, Prof. W.-Y., Shanghai Institute of Organic Chemistry, Chinese Academy of Sciences, 345 Linglin Lu, Shanghai 200032, China
(TEL: 313300) (21)

HUDNIK, Mrs. V., Kemijski Institut Boris Kidric, Hajdrihova 19, YU-61001 Ljubljana, Yugoslavia
(TEL: +38 (61) 263061) (37)

HULANICKI, Prof. A., Wydział Chemii, Uniwersytet Warszawski, ul. Pasteura 1, PL-02 093 Warszawa, Poland
(TEL: +48 (22) 221085. TELEX: 815439 UW PL) (27)

HUMERES, Prof. E., Departamento de Química, Universidade Federal de Santa Catarina, 88000 Florianópolis - SC, Brazil
(TEL: +55 (482) 33 0084/33 9219. TELEX: 482240 UFSC BR) (21)

HUNTER, Prof. R. J., School of Chemistry, University of Sydney, Sydney, New South Wales 2006, Australia
(TEL: +61 (2) 692 2176) (17)

HUYSKENS, Prof. P., Departement Scheikunde, Laboratorium voor Fysicochemie en Stralingschemie, Katholieke Universiteit te Leuven, Celestijnenlaan 200 F, B-3030 Heverlee, Belgium
(TEL: +32 (16) 20 06 56. TELEX: 23674 KULEUV B) (30)

IBRAHIM BIN ABDULLAH, Dr., Jabatan Kimia, Universiti Kebangsaan Malaysia, Bangi, Selangor, Malaysia
(TEL: +60 (3) 350001. TELEX: 31496 MA) (22)

IHNAT, Dr. M., Chemistry and Biology Research Institute, Agriculture Canada, Ottawa, Ontario K1A 0C6, Canada
(TELEX: 053 3283 CANAGRI CA) (38)

ILLERS, Dr. K. H., BASF Aktiengesellschaft, Mess- und Prüflabor, D-6700 Ludwigshafen/Rhein, Federal Republic of Germany
(TELEX: 464811 BASF D) (26)

ILLUMINATI, Prof. G., Dipartimento di Chimica Organica, Università di Roma "La Sapienza", Piazzale Aldo Moro 2, I-00185 Roma, Italy
(TEL: +39 (6) 4957808) (21, 22)

INAGAKI, Prof. M., Material Science, Toyohashi University of Technology, 1-1 Hibarigaoka, Tempaku-cho, Toyohashi, 440 Japan
(TEL: +81 (532) 47 0135) (20)

INCZÉDY, Prof. J., Analitikai Kémia Tanszék, Veszprémi Vegyipari Egyetem, Schönherz Z. u. 12, H-8201 Veszprém, Hungary
(TEL: +36 (80) 12550. TELEX: 32-397 VVEGYE H) (27)

INGMAN, Prof. F., Department of Analytical Chemistry, Royal Institute of Technology, Box, S-100 44 Stockholm 70, Sweden
(TEL: +46 (8) 7878183. TELEX: 10389 KTHB STOCKHOLM S) (27)

IONESCU, Prof. L. G., Instituto de Química, Universidade Federal do Rio Grande do Sul, 90000 Porto Alegre - RS, Brazil
(TEL: +55 (512) 36 9611. TELEX: (051) 1055) (17)

IRVING, Prof. H. M. N. H., Department of Analytical Science, University of Cape Town, Rondebosch 7700, Cape Province, Republic of South Africa
(TEL: +27 (21) 698531. TELEX: 57-21439 SA) (28)

ISE, Prof. N., Department of Polymer Chemistry, Faculty of Engineering, Kyoto University, Yoshida-machi, Sakyo-ku, Kyoto, 606 Japan
(TEL: + 81 (75) 751 2111. TELEX: 5422693 LIBKYU J) (23)

ISUYAMA, Dr. R., Instituto de Química, Universidade de São Paulo, Caixa Postal 20780, 01000 São Paulo - SP, Brazil
(TEL: +55 (11) 210 2122. TELEX: 011-22092 USPO BR) (11)

IWAMURA, Prof. H., Institute for Molecular Science, Myodaiji, Okazaki, Aichi-ken, 444 Japan
(TEL: +81 (564) 54 1111. TELEX: 4537-475 KOKKEN J) (22)

IYENGAR, Dr. V., Institut für Medizin, Kernforschungsanlage Jülich GmbH, Postfach 1913, D-5170 Jülich 1, Federal Republic of Germany
(TELEX: 833 556 KFA D) (38)

IZUTSU, Prof. K., Faculty of Science, Shinshu University, 3-1-1 Asahi, Matsumoto, 390 Japan
(TEL: +81 (263) 35 4600) (29)

JACKWERTH, Prof. E., Abteilung für Chemie, Ruhr-Universität Bochum, Postfach 102148, Universitätstrasse 150, D-4630 Bochum, Federal Republic of Germany
(TEL: +49 (234) 7006200) (27)

JACOBSBERG, Mlle. BRIGITTE, Tropical Products Sales SA, Unilever House, 51 Rue Montoyer, B-1040 Bruxelles, Belgium
(TEL: +32 (2) 516 75 11/523 20 80. TELEX: 21541 UNITEL BRU B) (33)

JAGANNATHAN, Dr. V., Biochemistry Division, National Chemical Laboratory, Pune-411008, India
(TEL: +91 (212) 56451. TELEX: 0145 266 NCL-PN IN) (32)

JÄGER, Dr. G., Bayer AG, D-5090 Leverkusen-Bayerwerk, Federal Republic of Germany
(TEL: +49 (214) 30 7483. TELEX: 85103 0 BY D) (10)

JAGNER, Dr. D., Department of Analytical and Marine Chemistry, University of Göteborg, S-412 96 Göteborg, Sweden
(TEL: +46 (31) 810100) (28)

JAIN, Prof. D. V. S., Department of Chemistry, Panjab University, Chandigarh-160014, India
(TEL: 31435) (15)

JAKABOS, Dr. A., Magyar Kémikusok Egyesülete, Anker Köz 1, H-1061 Budapest, Hungary
(TEL: +36 (1) 427 343) (5)

JAKOBI, Mr., Abteilung Hygiene und Arbeitsmedizin der Medizinischen Fakultät, RWTH Aachen, Lochnerstrasse 4-20, D-5100 Aachen, Federal Republic of Germany
(TEL: +49 (241) 8088880) (37)

JANÁK, Prof. J., Ústav Analytické Chemie, Československá Akademie Věd, Leninova ul. 82, CS-611 42 Brno, Czechoslovakia (27)

JANCSÓ, Dr. G., Max-Planck Institut für Chemie, Saarstrasse 23, D-6500 Mainz, Federal Republic of Germany
(TEL: +49 (6131) 305228. TELEX: 4187674 D) (30)

JANESCHITZ-KRIEGL, Prof. H., Physikalische Chemie, Johannes Kepler Universität Linz, Altenberger Strasse 69, A-4040 Linz, Austria
(TEL: +43 (732) 232381753) (24)

JANSSEN, Dr. P., Laboratoire IVP, Avenue Pierre Holoffe, B-1342 Limelette, Belgium (25)

JARDILLIER, Prof. J. C., Institut Jean Godinot, 1 rue du Général Koenig, F-51056 Reims Cédex, France
(TEL: +33 (26) 06 00 73/05 04) (35)

JARISCH, Dr. EDITH, Billrothstrasse 39, A-1190 Wien, Austria
(TEL: +43 (222) 361617) (11)

JÄRVISALO, Dr. J., Institute of Occupational Health, Arinatie 3, SF-00370 Helsinki 37, Finland
(TEL: +358 (0) 558961. TELEX: 121394 OCCUPHEALTH HELSINKI SF) (37)

JATLOW, Dr. P. I., Department of Laboratory Medicine, Yale University Medical School, POB 1001, New Haven, Connecticut 06504, USA
(TEL: +1 (203) 436 0854) (37)

JEANNIN, Prof. Y., Laboratoire de Chimie des Métaux de Transition, Université Pierre et Marie Curie, 4 place Jussieu, F-75230 Paris Cédex 05, France
(TEL: +33 (1) 336 25 25. TELEX: 200145 F) (10, 19)

JEDLIŃSKI, Prof. Z. J., Zakład Polimerów Polskiej Akademii Nauk, POB 49, ul. Marii Curie-Skłodowskiej 34, PL-41 800 Zabrze, Poland
(TEL: 710007) (23)

JEFFRIES, Dr. R., Shirley Institute, Didsbury, Manchester M20 8RX, UK
(TEL: +44 (61) 445 8141) (26)

JENCKS, Prof. W. P., Graduate Department of Biochemistry, Brandeis University, Waltham, Massachusetts 02254, USA
(TEL: +1 (617) 647 2726) (21)

JENKINS, Prof. A. D., School of Chemistry and Molecular Sciences, University of Sussex, Falmer, Brighton BN1 9QJ, UK
(TEL: +44 (273) 606755) (12, 23)

JENSEN, Dr. J. B., Fysisk-Kemisk Institut, Danmarks Tekniske Højskole, Bygning 206, DK-2800 Lyngby, Denmark
(TEL: +45 (2) 88 31 11. TELEX: 37529 DTHDIA DK) (30)

JEŻOWSKA-TRZEBIATOWSKA, Prof. BOGUSŁAWA, Instytut Chemii, Uniwersytet Wrocławski, ul. Joliot-Curie 14, PL-50 383 Wrocław, Poland
(TELEX: 0712791 PL) (17)

JIANG, Prof. X., Department of Physical Organic Chemistry, Shanghai Institute of Organic Chemistry, Chinese Academy of Sciences, 345 Linglin Lu, Shanghai 200032, China
(TEL: 313300) (21)

JOHANIDES, Prof. VERA, Faculty of Biotechnology, University of Zagreb, Pierottieva 6/IV, YU-41000 Zagreb, Yugoslavia
(TEL: +38 (41) 440422) (33)

JOHANSSON, Prof. G., Department of Analytical Chemistry, Chemical Center, University of Lund, Box 740, S-220 07 Lund 7, Sweden
(TEL: +46 (46) 108170. TELEX: S-33533 LUNIVER) (29)

JOHNSON, Prof. W. H., School of Physics and Astronomy, University of Minnesota, 116 Church Street SE, Minneapolis, Minnesota 55455, USA
(TEL: +1 (612) 373 5218) (19)

JOHNSON, Dr. A. R., Food Research Laboratory, Division of Food Research, Commonwealth Scientific and Industrial Research Organization, PO Box 52, North Ryde, New South Wales 2113, Australia
(TEL: +61 (887) 8333. TELEX: AA23407) (33)

JONES, Dr. R. A. Y., School of Chemical Sciences, University of East Anglia, University Plain, Norwich NR4 7TJ, UK
(TEL: +44 (603) 56161. TELEX: 97154 G) (21, 22)

JONES, Dr. R. N., "Ann Manor" Suite 601, 71 Somerset Street West, Ottawa, Ontario K2P 2G2, Canada
(TEL: +1 (613) 233 1565) (13, 17)

JORDAN, Prof. J., 152 Davey Laboratory, Department of Chemistry, Pennsylvania State University, University Park, Pennsylvania 16802, USA
(TEL: +1 (814) 865 2022. TELEX: 842 510) (29)

JORDANOV, Prof. N., Institute of General and Inorganic Chemistry, Bulgarian Academy of Sciences, Sofia 1113, Bulgaria (27)

JUHÁSZ, Dr. E., National Office of Measurements, Németvölgyi utca 37, H-1124 Budapest XII, Hungary
(TEL: +36 (1) 850 540. TELEX: 224856 OM H) (15)

JUILLARD, Dr. J., Laboratoire d'Etude des Interactions Solutés-Solvants, Université de Clermont 2, Les Cézeaux, BP 45, F-63170 Aubière, France
(TEL: +33 (16) 7326 41 10) (29)

JUSTICE, Dr. J. C., Laboratoire d'Electrochimie, Université Pierre et Marie Curie, 4 place Jussieu, F-75230 Paris Cédex 05, France
(TEL: +33 (1) 336 2525. TELEX: 200145 F) (16)

JUSZKIEWICZ, Prof. T., Zakład Farmakologii i Toksykologii, Instytut Weterynarii, Aleia Partyzantów 57, PL-24 100 Puławy, Poland
(TEL: 3051. TELEX: 0642401 IWET PL) (32)

KABANOV, Prof. V. A., Department of Macromolecular Chemistry, Faculty of Chemistry, Lomonosov Moscow State University, Leninskie Gory, SU-117234 Moscow, V-234, USSR
(TEL: +7 (095) 139 5583) (23, 24)

KADISH, Dr. K. M., Department of Chemistry, University of Houston, Central Campus, Houston, Texas 77004, USA
(TEL: +1 (713) 749 2639. TELEX: 76-2378 UH) (29)

KÅGEDAL, Dr. B., Klinisk-Kemiska Laboratoriet, Regionsjukhuset, Universitet i Linköping, S-581 85 Linköping, Sweden
(TEL: +46 (13) 193223) (36)

KAISER, Prof. E., Medizinisch-Chemisches Institut der Universität Wien, Währingerstrasse 10, A-1090 Wien IX, Austria
(TEL: +43 (222) 346442) (35)

KALIDAS, Prof. C., Department of Chemistry, Indian Institute of Technology, Madras-600036, India
(TEL: +91 (44) 415515/415342. TELEX: 041-7362 TECHMAS-MS IN) (30)

KALISZAN, Dr. R., Faculty of Pharmacy, Medical Academy, ul. Marksa 107, PL-80 416 Gdańsk, Poland (31)

KALLAY, Dr. N., Fizičko-Kemijski Zavod, Prirodoslovno-Matematičkog Fakulteta, Sveučilišta u Zagrebu, POB 163, Marulićev trg 19, YU-41001 Zagreb, Yugoslavia
(TEL: +38 (41) 446211/528) (15)

KALLNER, Dr. A., Klinisk-Kemiska Laboratoriet, Karolinska Sjukhuset, Box 60500, S-104 01 Stockholm 60, Sweden
(TEL: +46 (8) 7361943. TELEX: 17178 S) (35)

KALVIUS, Prof. G. M., Physik-Department E15, Fachbereich Physik, Technische Universität München, James Franck Strasse, D-8046 Garching bei München, Federal Republic of Germany
(TEL: +49 (89) 32092501. TELEX: 0522854 D) (6)

KALVODA, Dr. R., Ústav Fyzikální Chemie a Elektrochemie J Heyrovského, Československá Akademie Věd, Vlašská 9, CS-118 40 Praha 1-Malá Strana, Czechoslovakia
(TEL: +42 (2) 535132/531547-8) (29)

KAMACHI, Prof. M., Department of Polymer Chemistry, Faculty of Science, University of Osaka, 1-1 Machikaneyama, Toyonaka, Osaka, Japan (26)

KAMBANIS, Dr. S. M., Reichhold Chemicals Ltd., 1919 Wilson Avenue, Weston, Ontario M9M 1B1, Canada (25)

KAMBE, Prof. H., Faculty of Technology, Gunma University, Tenjin-cho, Kiryu, Gunma, 376 Japan (26)

KANNO, Prof. T., Department of Laboratory Medicine, Hamamatsu University School of Medicine, 3600 Handa-cho, Hamamatsu City, Shizuoka-ken, 431-31 Japan
(TEL: +81 (534) 35 2721) (36)

KAO, Prof. H., Department of Chemistry, Nanjing University, Nanjing, China (29)

KAPLAN, Prof. MARIA A., Instituto de Química, Universidade Federal Fluminense, Outeiro de São João Batista, s/n°, 24000 Niterói - RJ, Brazil
(TEL: +55 (21) 718 8590) (21)

KAPOOR, Prof. R. C., Department of Chemistry, University of Jodhpur, Jodhpur-342001, India
(TEL: 20740) (29)

KAPOULAS, Prof. V. M., Department of Biochemistry, University of Ioannina, Ioannina, Greece
(TEL: +30 (651) 30549) (33)

KAPPE, Prof. T., Institut für Organische Chemie der Universität Graz, Heinrichstrasse 28, A-8010 Graz, Austria (7)

KAPPLER, Dr. P., Centre de Recherches d'ATO, F-27470 Serquigny, France
(TEL: +33 (32) 44 10 06) (25)

KAPUR, Dr. B. M., Clinical Institute, Addiction Research Foundation, 33 Russell Street, Toronto, Ontario M5S 2S1, Canada
(TEL: +1 (416) 595 6181) (36)

KARASZ, Prof. F. E., Polymer Science and Engineering Department, University of Massachusetts, Amherst, Massachusetts 01003, USA
(TEL: +1 (413) 545 0433. TELEX: 95 5491) (26)

KASPRZAK, Dr. K. S., National Cancer Institute, Frederick Cancer Research Center, Building 538, Room 226, Frederick, Maryland 21701, USA
(TEL: +1 (301) 663 7659) (35, 36, 37)

KATO, Dr. M., Research Institute for Polymers and Textiles, 114 Yatabe-higashi, Tsukuba, Ibaraki, 305 Japan
(TEL: +81 (298) 54 6262) (24)

KATO, Dr. T., Central Research Laboratories, Ajinomoto Co. Inc., 1-1 Suzuki-cho, Kawasaki-ku, Kawasaki, 210 Japan
(TEL: +81 (44) 222 1111. TELEX: 22690 AJNOMOTO J) (32)

KAWANO, Prof. Y., Instituto de Química, Universidade São Paulo, Caixa Postal 20780, 01000 São Paulo - SP, Brazil
(TEL: +55 (11) 211 3120. TELEX: 011-22092 USPO BR) (17)

KEARNEY, Dr. P. C., Building 050, BARC-W, Pesticide Degradation Laboratory, Agricultural Environment Quality Institute, US Department of Agriculture, Beltsville, Maryland 20705, USA
(TEL: +1 (301) 344 3533) (13, 32, 34)

KEATING, Mr. P. B., Monsanto SA, Technical Center Europe, Rue Laid Burniat, B-1348 Louvain-la-Neuve, Belgium
(TEL: +32 (10) 41 81 11. TELEX: 59 056 MONTEC BO B) (24)

KEHIAIAN, Dr. H. V., Institut de Topologie et de Dynamique des Systèmes, Centre National de la Recherche Scientifique, Université de Paris VII, 1 rue Guy de la Brosse, F-75005 Paris, France
(TEL: +33 (1) 336 25 25) (16)

KELEN, Dr. T., Magyar Tudományos Akadémia, Központi Kémiai Kutató Intézete, POB 17, Pusztaszeri utca 59-67, H-1525 Budapest, Hungary
(TEL: +36 (1) 353 100. TELEX: 226686 KEMIA H) (24, 25)

KEMULA, Prof. W., Instytut Chemii Fizycznej, Polskiej Akademii Nauk, ul. Kasprzaka 44-52, PL-01 224 Warszawa, Poland
(TELEX: 817097 ICHF PL) (28)

KERR, Dr. G. T., Mobil Research and Development Corpn., Central Research Division, POB 1025, Princeton, New Jersey 08540, USA (7)

KERTES, Prof. A. S., Institute of Chemistry, Hebrew University of Jerusalem, Jerusalem 91904, Israel
(TEL: +972 (2) 585354. TELEX: 25391 HU IL) (18, 30)

KESTIN, Prof. J., Division of Engineering, Box D, Brown University, Providence, Rhode Island 02912, USA
(TEL: +1 (401) 863 2685. TELEX: 952-095 BRNTLXCTR-PVD) (16)

KETTRUP, Prof. A., Fachbereich 13, Universität Paderborn, Warburger Strasse 100, D-4790 Paderborn, Federal Republic of Germany (26)

KHANDELWAL, Dr. S., Industrial Toxicology Research Centre, POB 80, Mahatma Gandhi Marg, Lucknow-226001, India
(TEL: +91 (227) 33786) (37)

KHODAIR, Prof. A. I., Suez Canal University, Ismailia, Arab Republic of Egypt
(TEL: Cairo 749869. TELEX: 63297-SCU FM UN) (11)

KHODAKOVSKII, Prof. I. L., V I Vernadskii Institute of Geochemistry and Analytical Chemistry, Academy of Sciences of USSR, Kosygin Street 19, SU-117975, SZD-1, Moscow, V-334, USSR
(TEL: +7 (095) 137 5742) (16, 29, 30)

KHOO, Dr. K. H., Jabatan Kimia, Universiti Malaya, Lembah Pantai, Kuala Lumpur 22-11, Malaysia
(TEL: +60 (3) 555466. TELEX: 37453 MA) (30)

KIEFER, Prof. W., Physikalisches Institut, Universität Bayreuth, Postfach 3008, D-8580 Bayreuth, Federal Republic of Germany
(TEL: +49 (921) 552505) (17)

KIHLBORG, Prof. L., Department of Inorganic Chemistry, Arrhenius Laboratory, University of Stockholm, S-106 91 Stockholm, Sweden
(TEL: +46 (8) 162370. TELEX: 117 34 UNIVERS S) (20)

KING, Dr. W., State Laboratory, Abbotstown, Castleknock, Co. Dublin, Ireland
(TEL: +0044 (3531) 217820) (36)

KIRKBRIGHT, Prof. G. F., Department of Instrumentation and Analytical Science, University of Manchester Institute of Science and Technology, POB 88, Manchester M60 1QD, UK
(TEL: +44 (61) 236331. TELEX: 666094 G) (28)

KIRSCHNER, Prof. S., Department of Chemistry, Wayne State University, Detroit 2, Michigan 48202, USA (6)

KISAKÜREK, Dr. M. V., c/o Helvetica Chimica Acta, Faculté des Sciences, CH-1700 Fribourg, Switzerland (5)

KITAHARA, Prof. A., Department of Industrial Chemistry, Faculty of Engineering, Science University of Tokyo, Kagurasaka 1-3, Shinjuku-ku, Tokyo, 162 Japan
(TEL: +81 (3) 260 4271) (18)

KJAER, Prof. A., Organisk-Kemisk Laboratorium, Danmarks Tekniske Højskole, Bygning 201, DK-2800 Lyngby, Denmark
(TEL: +45 (2) 88 25 66. TELEX: 37529 DTHDIA DK) (10)

KLEIN, Dr. W., Fraunhofer-Institut für Toxikologie und Aerosolforschung, Institutsteil Grafschaft, D-5948 Schmallenberg, Federal Republic of Germany
(TEL: +49 (2972) 494-96. TELEX: 841520 D) (33)

KLESNEY, Mr. S. P., Central Report Index, 566 Building, Dow Chemical Co., Midland, Michigan 48640, USA
(TEL: +1 (517) 636 3754) (21)

KLINGENBERG, Prof. M., Institut für Physikalische Biochemie, Universität München, Goethestrasse 33, D-8000 München 2, Federal Republic of Germany
(TEL: +49 (89) 5996473) (13)

KLOCKOW, Prof. D., Abteilung für Chemie, Universität Dortmund, Postfach 500500, Otto Hahn Strasse, D-4600 Dortmund 50, Federal Republic of Germany
(TEL: +49 (231) 7553811) (28)

KLUMPP, Prof. G. W., Vakgroep Organische Chemie, Subfaculteit der Scheikunde, Vrije Universiteit, De Boelelaan 1083, 1081 HV Amsterdam, Netherlands
(TEL: +31 (20) 717451. TELEX: 11329 NL) (22)

KNOWLES, Dr. M. E., Food Science Laboratory, Ministry of Agriculture, Fisheries and Food, Haldin House, Queen Street, Norwich NR2 45X, UK
(TEL: +44 (603) 611712) (32)

KOCH, Dr. W., Research Department, Sandoz AG, CH-4000 Basel, Switzerland
(TEL: +41 (61) 24 31 19. TELEX: SD2B 62143 CH) (21)

KODAMA, Prof. Y., University of Occupational and Environmental Health, Japan School of Medicine, Honjo, Jahata Nishiku, Kitakyushu City, 807 Japan
(TEL: +81 (93) 603 1611) (37)

KOELMAN, Drs. F. H., Unichema Chemie BV, Postbus 2, 2800 AA Gouda, Netherlands
(TEL: +31 (1820) 12000. TELEX: 20661 NL) (33)

KOH, Dr. S. P., Jabatan Kimia, Universiti Malaya, Lembah Pantai, Kuala Lumpur 22-11, Malaysia
(TEL: +60 (3) 555466. TELEX: 37453 MA) (15)

KOHLER, Prof. F., Lehrstuhl für Thermodynamik, Ruhr-Universität Bochum, Postfach 102148, D-4630 Bochum, Federal Republic of Germany
(TEL: +49 (234) 7006389) (16)

KOJIMA, Prof. K., College of Environmental Health, Azabu University, 1-17-71 Fuchinobe, Sagamihara-shi, Kanagawa-ken, 229 Japan
(TEL: +81 (427) 54 7111) (32)

KOLINSKY, Dr., Ústav Makromolekulární Chemie, Československá Akademie Věd, Heyrovského náměstí 2, CS-162 06 Praha 616, Czechoslovakia (TEL: +42 (2) 353341. TELEX: 116/064 C) (25)

KOLLEK, Dr. H., Fraunhofer Institut für Angewandte Materialforschung, Lesumer Heerstrasse 36, D-2820 Bremen 77, Federal Republic of Germany (25)

KOMAREK, Prof. K. L., Institut für Anorganische Chemie der Universität Wien, Währingerstrasse 42, A-1090 Wien, Austria (TEL: +43 (222) 345424) (19, 20)

KON, Dr. H., Room B1-14, Building 2, Laboratory of Chemical Physics, NIADDK/National Institutes of Health, Bethesda, Maryland 20205, USA (TEL: +1 (301) 496 2667/1024) (17)

KOOISTRA, Ir. M. F., AKZO Research, PO Box 60, 6800 AB Arnhem, Netherlands (25)

KOPS, Prof. J., Instituttet for Kemiindustri, Danmarks Tekniske Højskole, Bygning 227, DK-2800 Lyngby, Denmark (TEL: +45 (2) 88 41 44. TELEX: 37529 DTHDIA DK) (23)

KOPTYUG, Prof. V. A., Institute of Organic Chemistry, Siberian Branch of Academy of Sciences of USSR, Prospekt Nauki 17, SU-630090 Novosibirsk 90, USSR (TEL: +7 (3832) 655 850) (10)

KORNHAUSER, Prof. ALEKSANDRA, Fakulteta za naravoslovje in tehnologijo, POB 18/1, Vegova 4, YU-61001 Ljubljana, Yugoslavia (TEL: +38 (61) 214326. TELEX: 31296 YU JOSTIN) (11)

KORDESCH, Prof. K. V., Technische Universität Graz, Stremayrgasse 16, A-8010 Graz, Austria (7)

KORYTA, Prof. J., Ústav Fyzikální Chemie a Elektrochemie J Heyrovského, Československá Akademie Věd, Opletalova 25, CS-110 00 Praha 1, Czechoslovakia (16)

KOTARSKI, Dr. A., Instytut Przemysłu Organicznego, ul. Annopol 6, PL-03 326 Warszawa, Poland (TEL: +48 (22) 112199. TELEX: 814844 IPO PL) (34)

KOTSOVASSILIS, Mr. C., c/o Association of Greek Chemists, 27 Kaningos Street, Athens 106 82, Greece (36)

KOVACICOVA, Dr. J., Výskumný Ústav Potravinárskí, Trenčianska ul. 53, CS-825 08 Bratislava, Czechoslovakia (TEL: +42 (7) 9761355) (34)

KOWALSKI, Prof. B. R., Laboratory for Chemometrics, Department of Chemistry, BG-10, University of Washington, Seattle, Washington 98195, USA (28)

KRATOCHVÍL, Dr. P., Ústav Makromolekulární Chemie, Československá Akademie Věd, Heyrovského náměstí 2, CS-162 06 Praha 616, Czechoslovakia (TEL: +42 (2) 360341. TELEX: 122019 IMCP C) (23)

KRAUS, Prof. G. A., Department of Chemistry, Iowa State University, Ames, Iowa 50011, USA (TEL: +1 (515) 294 4633. TELEX: 910 520 1167) (22)

KRAUSE, Dr. P., Bundesgesundheitsamt, Corrensplatz 1, D-1000 Berlin 33, Federal Republic of Germany (TEL: +49 (30) 83082715/2420) (37)

KRETZSCHMAR, Dr. H. J., Bundesanstalt für Materialprüfung (BAM), Unter den Eichen 87, D-1000 Berlin 45 (Dahlem), Federal Republic of Germany (TEL: +49 (30) 81043140. TELEX: 183261 BAMB D) (25)

KROGH, Prof. P., Department of Microbiology, Royal Dental College, Juliane Mariesvej 30^2, DK-2100 København Ø, Denmark (TEL: +45 (1) 39 37 07) (32)

KRÖNERT, Prof. W., Max-von-Pettenkofer Institut des Bundesgesundheitsamtes, Postfach 330013, Thielallee 88-92, D-1000 Berlin 33, Federal Republic of Germany (TEL: +49 (30) 83082463. TELEX: 0184016 D) (32)

KRYSZEWSKI, Prof. M., Centrum Badań Molekularnych i Makromolekularnych, Polska Akademii Nauk, Boczna 5, PL-90 362 Łódź, Poland (TEL: +48 (4) 48981. TELEX: 886742 PAN PL) (23)

KUBACKI, Dr. S. J., HFF 459, FOB 8, Room 4880, Food and Drug Administration, US Department of Health and Human Services, 200 C Street SW, Washington, DC 20204, USA (TEL: +1 (202) 245 1411. TELEX: 7108229530) (32)

KUBOUCHI, Dr. Y., Daikin Industries Ltd., 1-12-39 Umeda, Kita-ku, Osaka, 530 Japan (TEL: +81 (6) 346 1201. TELEX: 523 3844 J) (24)

KUCHITSU, Prof. K., Department of Chemistry, Faculty of Science, University of Tokyo, 3-1 Hongo 7-chome, Bunkyo-ku, Tokyo, 113 Japan (TEL: +81 (3) 812 2111. TELEX: 25510 UNITOKYO J) (12, 15)

KÜHN, Dr. K., Pfarrer-Jägers-Strasse 4, D-5352 Zülpich-Ülpenich, Federal Republic of Germany (TEL: +49 (2252) 3414) (14, 36, 37)

KULIN, Mr. L. J., Polymergruppen, Chalmers Tekniska Högskola, Göteborg Universitet, S-412 96 Göteborg, Sweden (TEL: +46 (31) 810100. TELEX: 2369 CHERLBIB S) (25)

KÜNZEL, Dr. W., Amt für Standardisierung, Messwesen und Warenprüfung (ASMW), Fürstenwalder Damm 388, DDR-1162 Berlin, German Democratic Republic (TEL: +37 (2) 6702891. TELEX: 112630 ASMW DD) (15)

KURIACOSE, Prof. J. C., Department of Chemistry, Indian Institute of Technology, Madras-600036, India (TEL: +91 (44) 415342. TELEX: 041-7362 TECHMAS-MS IN) (18)

KURIHARA, Dr. N., Radioisotope Research Center, Kyoto University, Yoshida Konoecho, Sakyo-ku, Kyoto, 606 Japan (TEL: +81 (75) 751 2111. TELEX: 5422693 LIBKYU J) (33)

KURUCZ, Dr. ÉVA, Növényolajipari és Mosószergyártó Vállalat, Horváth Mihály utca 2, H-1151 Budapest, Hungary (TEL: +36 (1) 690154. TELEX: 5366 BP H) (32, 33)

KUZMIN, Prof. M. G., Laboratory of Photochemistry, Faculty of Chemistry, Lomonosov Moscow State University, Leninskie Gory, SU-117234 Moscow, V-234, USSR (TEL: +7 (095) 139 2484) (22)

VAN LAAR, Dr. J. A. W., De Genestelaan 32, Bilthoven, Netherlands (25)

L'ABBE, Prof. G., Laboratorium voor Macromoleculaire en Organische Scheikunde, Katholieke Universiteit te Leuven, Celestijnenlaan 200 F, B-3030 Heverlee, Belgium (TEL: +32 (16) 20 06 56. TELEX: 23674 KULEUV B) (21)

DE LAETER, Dr. J. R., School of Physics and Geosciences, Western Australian Institute of Technology, Kent Street, Bentley, Western Australia 6102, Australia (TEL: +61 (9) 350 7093. TELEX: AA92983) (19)

LAGO, Dr. REGINA C. A., Centro de Tecnologia Agricola e Alimentar - EMBRAPA, Rua Jardim Botânico 1024, 22460 Rio de Janeiro - RJ, Brazil (TEL: +55 (21) 239 6290. TELEX: (021) 33267 EBPA BR) (33)

LAGOWSKI, Prof. J. J., Department of Chemistry, University of Texas at Austin, Austin, Texas 78712, USA (TEL: +1 (512) 471 3288) (12)

LAITINEN, Dr. R., Department of Chemistry, Helsinki University of Technology, Otaniemi, Otakaari 1, SF-02150 Espoo 15, Finland (TEL: +358 (0) 4512792. TELEX: 125161 HTKK SF) (19)

LAMBERT, Dr. I., Département de Recherche et Analyse, Centre d'Etudes Nucléaires de Saclay, F-91191 Gif-sur-Yvette, France (31)

LAMOTTE, Dr. A., Service Central d'Analyse du Centre National de la Recherche Scientifique, Exchangeur de Solaize, BP 22, F-69390 Vernaison, France
(TELEX: 900225 CNRS VIRLB F) (27, 28)

LANDAAS, Dr. S., Sentrallaboratoriet, Ullevål Sykehus, Oslo 1, Norway (35)

LANDOLT, Prof. D., Département des Matériaux, Laboratoire de Métallurgie Chimique, Ecole Polytechnique Fédérale de Lausanne, 34 chemin de Bellerive, CH-1007 Lausanne, Switzerland
(TEL: +41 (21) 47 29 81. TELEX: 24478 CH) (16)

LANGE, Dr. H., Bayer AG, FE-DPP-Verfahrensanalytik, B 211, D-5090 Leverkusen-Bayerwerk, Federal Republic of Germany
(TEL: +49 (214) 3031471. TELEX: 85103 0 BY D) (23)

LAQUA, Prof. K., Institut für Spektrochemie und Angewandte Spektroskopie, Postfach 778, Bunsen-Kirchhoff-Strasse 11, D-4600 Dortmund 1, Federal Republic of Germany
(TEL: +49 (231) 129001) (28)

LARA, Dr. W., Instituto Adolfo Lutz, Avenida Dr. Arnaldo 355, 01246 São Paulo - SP, Brazil (34)

LAUN, Dr. H. M., BASF Aktiengesellschaft, D-ZKM-G201, D-6700 Ludwigshafen/Rhein, Federal Republic of Germany
(TEL: +49 (621) 6022068. TELEX: 464811 BASF D) (24)

LAUWERYS, Prof. R., Unité de Toxicologie Industrielle et Médical, Université Catholique de Louvain, Boite 30-54, Clos Chapelle-aux-Champs 30, B-1200 Bruxelles, Belgium
(TEL: +32 (2) 762 34 00) (37)

LAZAROV, Prof. D. L., Department of Chemical Education, Chemical Faculty of Sofia University, Anton Ivanov Str. 1, Sofia 1126, Bulgaria (11)

LEARY, Dr. G. J., Department of Scientific and Industrial Research, Chemistry Division, Private Bag, Petone, New Zealand
(TEL: +64 (4) 666919. TELEX: 3814 PHYSICS NZ) (24)

LEHMANN, Dr. H. P., Department of Pathology, Louisiana State University Medical Center, 1901 Perdido Street, New Orleans, Louisiana 70112, USA
(TEL: +1 (504) 568 6057) (36)

LEICHNITZ, Dipl.-Ing. K., Drägerwerk AG, Hauptabteilung Chemische Gasmesstechnik, Postfach 1339, Moislinger Allee 53-55, D-2400 Lübeck 1, Federal Republic of Germany
(TEL: +49 (451) 8822335. TELEX: 026807 D) (32)

LEIGH, Dr. G. J., Agricultural Research Council, Unit of Nitrogen Fixation, University of Sussex, Brighton BN1 9RQ, UK
(TEL: +44 (273) 606755) (19)

LEMPERT, Prof. K., Szerves Kémiai Tanszék, Budapesti Müszaki Egyetem, Gellért tér 4, H-1521 Budapest, Hungary
(TEL: +36 (1) 452 178. TELEX: 225931 MÜEGY H) (22)

LE NEINDRE, Prof. B., Laboratoire des Interactions Moléculaires et des Hautes Pressions du Centre National de la Recherche Scientifique, Centre Universitaire Paris Nord, avenue Jean-Baptiste Clément, F-93430 Villetaneuse, France
(TEL: +33 (1) 821 61 70. TELEX: 204135 LABOBEL F) (15)

LEOLINI, Dr. C., Centro Ricerche "G. Donegani", Montedison SpA, Via Fausler 1, I-28100 Novara, Italy (25)

LEONARD, Dr. M. A., Department of Chemistry, David Keir Building, Queen's University of Belfast, Stranmillis Road, Belfast BT9 5AG, Northern Ireland, UK
(TEL: +0044 (0232) 661111. TELEX: 74487 G) (28)

LEONARD, Prof. N. J., 470B Roger Adams Laboratory, Box 42, School of Chemical Sciences, University of Illinois at Urbana-Champaign, 1209 West California Street, Urbana, Illinois 61801, USA
(TEL: +1 (217) 333 0363. TELEX: 910-245-2434 TWK) (11, 21)

LEUNG, Dr. F. Y., Department of Clinical Biochemistry, University Hospital, POB 5339, Postal Station A, London, Ontario N6A 5A5, Canada
(TEL: +1 (519) 673 6500) (38)

LEVANDER, Dr. O. A., Nutrition Institute, US Department of Agriculture, Beltsville, Maryland 20705, USA (38)

LEVER, Prof. A. B. P., Department of Chemistry, Faculty of Science, York University, 4700 Keele Street, Downsview, Ontario M3J 1P3, Canada
(TEL: +1 (416) 667 2309. TELEX: 065 24736 CA) (19)

LEVIN, Dr. Ö., Margarinbolaget AB, Box 30025, S-104 25 Stockholm 30, Sweden
(TEL: +46 (8) 541360. TELEX: 19571 MARG S) (33)

LEVY, Prof. M., Department of Plastics Research, Weizmann Institute of Science, Rehovot 76100, Israel
(TELEX: 361900 WIX IL) (24)

LEWIN, Prof. M., Israel Fiber Institute, POB 8001, Jerusalem, Israel (23)

LEWIS, Dr. F. A., Department of Chemistry, David Keir Building, Queen's University of Belfast, Stranmillis Road, Belfast BT9 5AG, Northern Ireland, UK
(TEL: +0044 (0232) 661111. TELEX: 74487 G) (30)

LEYDEN, Dr. D., Chemistry Department, Colorado State University, Fort Collins, Colorado 80521, USA (28)

L'HER, Dr. M., Laboratoire de Chimie Analytique, Université de Bretagne, 6 avenue le Gorgeu, F-29283 Brest Cédex, France
(TEL: +33 (98) 03 16 94) (29)

LI, Dr. Z., Research Institute of Elemento-Organic Chemistry, Nankai University, Tianjin, China (34)

LIDE, Jr., Dr. D. R., Office of Standard Reference Data, A323 Physics Building, National Bureau of Standards, US Department of Commerce, Washington, DC 20234, USA
(TEL: +1 (301) 921 2467. TELEX: 898493 GARG) (10, 15)

LIÉBECQ, Prof. C., Boulevard de la Constitution 69/054, B-4020 Liège, Belgium
(TEL: +32 (41) 42 76 44) (13)

LIM, Dr. C. L., School of Applied Sciences, Universiti Sains Malaysia, Penang, Malaysia (32)

LIND, Mr. B., Department of Environmental Hygiene, Karolinska Institute, Box 60400, S-104 01 Stockholm 60, Sweden
(TEL: +46 (8) 236900. TELEX: 19530 KI ENV S) (37)

LINDBERG, Prof. J. J., Department of Wood and Polymer Chemistry, University of Helsinki, Meritullinkatu 1 A, SF-00170 Helsinki 17, Finland
(TEL: +358 (0) 650211) (26)

LINDER, Prof. P. W., Department of Physical Chemistry, University of Cape Town, Rondebosch 7700, Cape Province, Republic of South Africa
(TEL: +27 (21) 698531. TELEX: 57-21439 SA) (30)

LINDINGER, Prof. W., Institut für Atomphysik, Karl Schönherr-Strasse 3, A-6020 Innsbruck, Austria
(TEL: +43 (5222) 33601) (15)

LINDSTEDT, Prof. S., Department of Clinical Chemistry, Sahlgren's Hospital, S-413 45 Göteborg, Sweden
(TEL: +46 (31) 602426) (35)

LINKO, Prof. M., Biotechnical Laboratory, Technical Research Centre of Finland, Tietotie 2, SF-02150 Espoo 15, Finland
(TEL: +358 (0) 4565150. TELEX: 122972 VTT HA SF) (32)

LINKO, Prof. R., Department of Chemistry and Biochemistry, University of Turku, Vesilinnantie 5, SF-20500 Turku 50, Finland
(TEL: +358 (21) 645748) (33)

LITTLER, Dr. J. S., School of Chemistry, University of Bristol, Cantock's Close, Bristol BS8 1TS, UK
(TEL: +44 (272) 24161. TELEX: 44174 BUPHYS G) (21)

LLAURO, Mme., Laboratoire de Matériaux Organiques du Centre National de la Recherche Scientifique, BP 24, F-69390 Vernaison, France
(TEL: +33 (7) 802 13 44. TELEX: 900225 CNRS VILRB F) (25)

LOBO, Prof. V. M. M., Department of Chemistry, University of Coimbra, P-3000 Coimbra, Portugal (16)

LODGE, Prof. A. S., Rheology Research Center, 1500 Johnson Drive, Madison, Wisconsin 53706, USA
(TEL: +1 (608) 263 1571. TELEX: 26 54 52) (24, 25)

LODI, Prof. F., Istituto di Medicina Legale e delle Assicurazioni Sezione Tossicologie, Forense Università, Via Mangiagalli 37, I-20133 Milano, Italy (36)

LOENING, Dr. K. L., Chemical Abstracts Service, POB 3012, 2540 Olentangy River Road, Columbus, Ohio 43210, USA
(TEL: +1 (614) 421 3633. TELEX: 810 482 1608 CHEM ABS COL) (11, 12, 13, 24)

LOKE, Prof. SU ENG, Jabatan Kimia, Universiti Malaya, Lembah Pantai, Kuala Lumpur 22-11, Malaysia
(TEL: +60 (3) 555466. TELEX: 37454 MA) (12)

LORIMER, Prof. J. W., Department of Chemistry, University of Western Ontario, London, Ontario N6A 5B7, Canada
(TEL: +1 (519) 679 2393) (27, 30)

LOUS, Prof. P., Department of Clinical Chemistry, Bispebjerg Hospital, Bispebjerg Bakke 23, DK-2400 København NV, Denmark
(TEL: +45 (1) 81 12 50) (37)

LOVEGROVE, Mr. D. H. W., Carrs Paints Ltd., Westminster Works, Alvechurch Road, West Heath, Birmingham B31 3PG, UK (25)

LOZAC'H, Prof. N., Institut des Sciences de la Matière et du Rayonnement, Université de Caen, 5 avenue d'Edimbourg, F-14032 Caen Cédex, France
(TEL: +33 (31) 93 37 14) (21)

LU, Prof. J., Fujian Institute of Research on the Structure of Matter, Chinese Academy of Sciences, Fuzhon, Fujian, China (15, 17)

LUCA, Prof. C., Institutul de Cercetări Chimice (ICECHIM), 202 Splaiul Independenţei, sect. 6, R-77208 Bucureşti, Romania
(TELEX: 10944 ICECHIM R) (29)

LÜCK, Dr. E., Hoechst AG, Forschung und Entwicklung Lebensmitteltechnik, D 729, Postfach 800320, D-6230 Frankfurt/Main 80, Federal Republic of Germany
(TEL: +49 (611) 3055542. TELEX: 41234 HOEAG D) (32)

LUTHARDT, Dr. H. J., Deutsche AKZO Coatings GmbH, Postfach 300709, D-7000 Stuttgart 30, Federal Republic of Germany (25)

LUXON, Mr. S. G., Maidensgrove, Riversdale, Bourne End, Buckinghamshire SL8 5RB, UK
(TEL: +44 (6285) 27261) (12, 32)

LUZ, Prof. Z., Department of Isotope Research, Weizmann Institute of Science, Rehovot 76100, Israel
(TELEX: 361900 WIX IL) (17)

L'VOV, Prof. B. V., Physico-Mathematical Department, Polytechnical Institute, Polytechnical Street 19, SU-195251 Leningrad, USSR
(TEL: +7 (812) 221 8982) (28)

LYKLEMA, Prof. J., Laboratorium voor Fysische en Kolloidchemie, Landbouwhogeschool, De Dreyen 6, 6703 BC Wageningen, Netherlands
(TEL: +31 (8370) 82279) (18)

MacLEAN, Dr. D. B., Association of Official Analytical Chemists, Suite 210, 1111 North Nineteenth Street, Arlington, Virginia 22209, USA (5)

McBRYDE, Prof. W. A. E., Department of Chemistry, University of Waterloo, Waterloo, Ontario N2L 3G1, Canada
(TEL: +1 (519) 885 1211. TELEX: 06955259 CA) (29, 30)

McCARTHY, Dr. MARY, Biocon Ltd., Kilnageary, Carrigaline, Co. Cork, Ireland (33)

McCARTNEY, Dr. E. R., Department of Ceramic Engineering, University of New South Wales, PO Box 1, Kensington, New South Wales 2033, Australia
(TEL: +61 (2) 662 2420. TELEX: AA26054) (20)

McENANEY, Dr. B., School of Materials Science, University of Bath, Claverton Down, Bath BA2 7AY, UK
(TEL: +44 (225) 61244. TELEX: 449097 G) (20)

McEWAN, Dr. JOAN C., Special Studies and Services Branch, Ontario Ministry of Labour, 400 University Avenue, 8th Floor, Toronto, Ontario M7A 1T7, Canada
(TEL: +1 (416) 965 6375) (37)

McLACHLAN, Dr. D. R. C., Department of Physiology, University of Toronto, Ontario M5S 1A8, Canada
(TEL: +1 (416) 978 2011) (38)

McLENNAN, Prof. D. J., Department of Chemistry, University of Auckland, Private Bag, Auckland, New Zealand
(TEL: +64 (9) 737999. TELEX: 21480 NZ) (22)

McNAUGHT, Dr. A. D., Royal Society of Chemistry, Burlington House, Piccadilly, London W1V 0BN, UK
(TEL: +44 (1) 734 9864. TELEX: 268001 G) (21)

MACHATA, Prof. G., Institut für Gerichtliche Medizin der Universität, Sensengasse 2, A-1080 Wien, Austria
(TEL: +43 (222) 426278/9) (37)

MACHO, Dr. V., Výskumný Ústav Pre Petrochémiu, CS-972 71 Nováky, Czechoslovakia
(TEL: 925. TELEX: 72343 VUPET C) (10)

MACOSKO, Prof. C., Chemical Engineering and Materials Science, 151 Chemical Engineering, University of Minnesota, 421 Washington Avenue SE, Minneapolis, Minnesota 55455, USA
(TEL: +1 (612) 373 4815) (24)

MACZYNSKA, Dr. ZOFIA, Instytut Chemii Fizycznej, Polskiej Akademii Nauk, ul. Kasprzaka 44-52, PL-01 224 Warszawa, Poland
(TEL: +48 (22) 328392. TELEX: 817097 ICHF PL) (30)

MACZYNSKI, Dr. A., Instytut Chemii Fizycznej, Polskiej Akademii Nauk, ul. Kasprzaka 44-52, PL-01 224 Warszawa, Poland
(TEL: +48 (22) 328392. TELEX: 817097 ICHF PL) (30)

MADDAMS, Dr., BP Research Centre, Chertsey Road, Sunbury on Thames, Middlesex TW16 7LN, UK
(TEL: +44 (9327) 81234. TELEX: 929732 G) (25)

MAGNÉLI, Prof. A., Arrhenius Laboratory, University of Stockholm, S-106 91 Stockholm, Sweden
(TEL: +46 (8) 611969/162368. TELEX: 117 34 UNIVERS S) (20)

MAJNUSZ, Dr. J., Institute of Physical Chemistry and Technology of Polymers, Silesian Technical University, ul. Kuczewskiego 9, PL-44 100 Gliwice, Poland
(TEL: +48 (32) 314564. TELEX: 036104 PL) (24)

MALANOWSKI, Dr. S., Instytut Chemii Fizycznej, Polskiej Akademii Nauk, ul. Kaszprzaka 44-52, PL-01 224 Warszawa, Poland
(TEL: +48 (22) 326713. TELEX: 817097 ICHF PL) (16)

MANDEL, Prof. M., Gorlaeus Laboratoria der Rijksuniversiteit, Postbus 9502,
2300 RA Leiden, Netherlands
(TEL: +31 (71) 148333) (11, 23)

MANDELBAUM, Prof. A., Department of Chemistry, Technion-Israel Institute
of Technology, Technion City, Haifa 32000, Israel
(TEL: +927 (4) 225111. TELEX: 46650 TECLI IL) (17)

MANLEY, Dr. T. R., Department of Materials Science, Newcastle upon Tyne Polytechnic,
Ellison Building, Newcastle upon Tyne NE1 8ST, UK (26)

MANNE, Prof. R., Department of Chemistry, University of Bergen, Allégate 41,
N-5014 Bergen-Universitet, Norway
(TEL: +47 (5) 320040. TELEX: 42877 N) (28)

MANNING, Dr. R. G., Durkee Foods Division, SCM Corpn., Strongsville,
Ohio 44136, USA (33)

MANO, Prof. ELOISA B., Instituto de Macromoléculas - IMA, Universidade Federal
do Rio de Janeiro, Caixa Postal 68 525, 21944 Rio de Janeiro - RJ, Brazil
(TEL: +55 (21) 270 1035. TELEX: 2122924 UFRJ BR) (23, 24)

MARCH, Prof. J., Department of Chemistry, Adelphi University, Garden City,
Long Island, New York 11530, USA
(TEL: +1 (516) 294 8700. TELEX: 5102226817) (21, 22)

MARCHESSAULT, Dr. R. H., Xerox Research Centre of Canada, 2480 Dunwin Drive,
Mississauga, Ontario L5L 1J9, Canada
(TEL: +1 (416) 828 6200) (10, 24)

MARCHIDAN, Dr. D., Center of Physical Chemistry, Institutul de Cercetări Chimice
(ICECHIM), 202 Splaiul Independenţei, R-77208 Bucureşti, Romania
(TEL: 496675. TELEX: 11171 B MCHIM R) (16)

MARCUS, Prof. Y., Department of Inorganic and Analytical Chemistry,
Hebrew University of Jerusalem, Jerusalem 91904, Israel
(TEL: +972 (2) 585341. TELEX: 25391 HU IL) (27, 29)

MARCZENKO, Prof. Z., Instytut Chemii Ogólnej i Technologii Nieorganicznej
Politechniki Warszawskiej, ul. Noakowskiego 3, PL-00 664 Warszawa, Poland
(TEL: +48 (22) 283 339) (27)

MARIČIĆ, Dr. S., Nacionalna i Sveučilišna Biblioteka, POB 550, Marulićev trg 21,
YU-41001 Zagreb, Yugoslavia
(TEL: +38 (41) 446123. TELEX: 22206 BICH YU) (11)

MAROTO, Dr. A., Departamento Química de Reactores, Comisión Nacional de Energía
Atómica, Avenida del Libertador 8250, 1429 Buenos Aires, Argentina
(TEL: +54 (1) 707711. TELEX: 21388 PREAT AR) (17)

MAROV, Prof. I. N., V I Vernadskii Institute of Geochemistry and
Analytical Chemistry, Academy of Sciences of USSR, Kosygin Street 19,
SU-117975, SZD-1, Moscow, V-334, USSR (29)

MARSEL, Prof. J., Institut za Nuklearne Nauke Jožef Stefan, Jamova 31,
YU-61001 Ljubljana, Yugoslavia
(TEL: +38 (61) 263261) (17)

MARSH, Prof. H., Northern Coke Laboratory, School of Chemistry,
University of Newcastle upon Tyne, Newcastle upon Tyne NE1 7RU, UK
(TEL: +44 (632) 328511. TELEX: 53654 UNINEW G) (20)

MARSH, Dr. K. N., Thermodynamics Research Center, Texas A & M University,
College Station, Texas 77843, USA
(TEL: +1 (409) 845 4995. TELEX: 510 892 7689) (15, 16)

MARTI, Dr. E., CIBA-GEIGY AG, CH-4002 Basel, Switzerland
(TELEX: 62355 CH) (26)

MARTIN, Prof. R. F., Pathology Department, Methodist Hospital, 6565 Fannin,
Houston, Texas 77030, USA (28)

MARTIN, Prof. R. L., Monash University, Wellington Road, Clayton, Victoria 3168, Australia
(TEL: +61 (3) 541 2000. TELEX: AA32691 MONASH) (19)

MARVEL, Dr. J. T., Research Division, Monsanto Agricultural Products Co., 800 North Lindbergh Boulevard, St. Louis, Missouri 63166, USA
(TEL: +1 (314) 694 5108. TELEX: 44-7282) (21)

MASON, Mr. D. M., Institute of Gas Technology, 3424 South State Street, Chicago, Illinois 60616, USA (30)

MASON, Dr. G. W., Ivon Watkins-Dow Ltd., POB 144, New Plymouth, New Zealand (34)

MASON, Dr. S. G., Pulp and Paper Research Institute of Canada, Department of Chemistry, Pulp and Paper Building, McGill University, 3420 University Street, Montreal, Québec H3A 2A7, Canada (18)

MASSABNI, Prof. A. C., Instituto de Química - UNESP, Caixa Postal 174, 14800 Araraquara - SP, Brazil (19)

MASSART, Prof. D. L., Farmaceutisch Instituut, Vrije Universiteit Brussel, Laarbeeklaan 103, B-1090 Brussel, Belgium
(TEL: +32 (2) 478 15 20) (27)

MASSON, Dr. MARY, Department of Chemistry, University of Aberdeen, Old Aberdeen AB9 2UE, UK (30)

MASTROMATTEO, Dr. E., Medical Department, International Nickel Co. Ltd., 1 First Canadian Place, Toronto, Ontario M5X 1C4, Canada
(TEL: +1 (416) 361 7869) (37)

MASUDA, Dr. T., Department of Polymer Chemistry, Faculty of Engineering, Kyoto University, Yoshida-machi, Sakyo-ku, Kyoto, 606 Japan
(TEL: +81 (75) 751 2111. TELEX: 5422693 LIBKYU J) (24)

MATHIEU, Prof. J., Centre de Recherches Roussel Uclaf SA, BP 9, 102 route de Noisy, F-93230 Romainville, France
(TEL: +33 (1) 843 93 10. TELEX: 680501 ROUSSEL RMVI F) (21)

MATHOT, Dr. V. B. F., Department of Fundamental Polymer Research, Dutch State Mines, PO Box 18, 6160 MD Geleen, Netherlands
(TELEX: 56727 NL) (26)

MATOUSEK, Prof. J., Katedra Technologie Silikátů, Vysoká Škola Chemicko-Technologická v Praze, Suchbátarova 5, CS-166 28 Praha 6-Dejvice, Czechoslovakia
(TEL: +42 (2) 320259. TELEX: 122744 VSCH C) (20)

MATTIELLO, Dr. G., Laboratorio Provinciale d'Igiene e Profilassi, Sezione Chimica, Campo della Lana 601, I-30125 Venezia, Italy
(TEL: +39 (41) 26748) (37)

MATTSSON, Prof. E., Swedish Corrosion Institute, Box 5607, S-114 86 Stockholm, Sweden
(TEL: +46 (8) 222540. TELEX: 14375 STURES) (16)

MATUSIEWICZ, Dr. H., Department of Chemistry, University of Massachusetts, Amherst, Massachusetts 01003, USA
(TEL: +1 (413) 545 2294) (38)

MAURAS, Dr. Y., Laboratoire de Pharmacologie, C.H.U., F-49036 Angiers, France
(TEL: +33 (41) 88 69 51) (38)

MAURER, Mr. J. J., Exxon Research and Engineering Co., Corporate Research, Linden, New Jersey 07036, USA (26)

MEDVEDEV, Dr. V. A., Institute for High Temperatures, Academy of Sciences of USSR, Korovinskoye Road, SU-127412 Moscow, USSR
(TEL: +7 (095) 232 4205) (15)

MEHROTRA, Prof. R. C., Department of Chemistry, University of Rajasthan,
Jaipur-302004, India
(TEL: 60088/60271) (19)

MEHTA, Prof. G., School of Chemistry, Indian Institute of Technology,
Kanpur-208016, India (21)

MEININGER, Dr. J., Société Métallurgique le Nickel, 33 avenue du Maine,
F-75755 Paris Cédex 15, France
(TEL: +33 (1) 538 3196/6868) (37)

MEISSNER, Prof. J., Technisch-Chemisches Laboratorium der Eidgenössischen
Technischen Hochschule Zürich, Universitätstrasse 6, CH-8092 Zürich, Switzerland
(TEL: +41 (1) 256 45 40. TELEX: 53178 ETHBI CH) (24, 25)

MELHUISH, Dr. W. H., Institute of Nuclear Sciences, Department of Scientific
and Industrial Research, Private Bag, Lower Hutt, New Zealand
(TEL: +64 (4) 699199. TELEX: 3814 PHYSICS NZ) (28)

MEMMING, Dr. R., Forschungslaboratorium, Philips GmbH, Vogt-Kolin-Strasse 30,
D-2000 Hamburg 54, Federal Republic of Germany
(TEL: +49 (40) 5493) (16)

MERMET, Dr. J. M., Service Central d'Analyse du Centre National de la Recherche
Scientifique, BP 22, F-69390 Vernaison, France
(TEL: +33 (7) 802 13 44. TELEX: 900225 CNRS VIRLB F) (28)

METSELAAR, Prof. R., Laboratorium voor Fysische Chemie, Technische Hogeschool
Eindhoven, Postbus 513, 5600 MB Eindhoven, Netherlands
(TEL: +31 (40) 473054/473044. TELEX: 51163 NL) (20)

VON MEYENBURG, Prof. K., Department of Microbiology, Danmarks Tekniske Højskole,
Bygning 221, DK-2800 Lyngby, Denmark
(TEL: +45 (2) 88 40 66. TELEX: 37529 DTHDIA DK) (32)

MEYER, Dr. H. H., Chemische Werke Hüls AG, ZB Forschung und Entwicklung,
Postfach 1320, D-4370 Marl-Hüls 1, Federal Republic of Germany
(TEL: +49 (2365) 495210. TELEX: 829211-0 CW D) (24)

MEYER, Prof. K., Institut für Chemische Technologie, Akademie der Wissenschaften
der DDR, Rudower Chaussee 5, DDR-1199 Berlin-Adlershof, German Democratic Republic
(TEL: +37 (2) 6743246. TELEX: 112537 ADWAD DD) (18)

MEYER, Dr. R. A., Ivan A. Gettinge Laboratories, Carbon and Polymer Department,
Material Science, Aerospace Corpn., POB 92957, Los Angeles, California 90009, USA
(TEL: +1 (213) 648 5000) (20)

MEYERHOFF, Prof. G., Institut für Physikalische Chemie, Johannes-Gutenberg
Universität Mainz, D-6500 Mainz, Federal Republic of Germany
(TEL: +49 (6131) 392330. TELEX: 4187476 UNI D) (26)

MEYERINK, Mr. N. L. J., Central Laboratory, Dutch State Mines,
PO Box 18, 6160 MD Geleen, Netherlands
(TEL: +31 (4494) 69111. TELEX: 36027 NL) (25)

MICHAEL, Mr. T. H. G., Chemical Institute of Canada, 151 Slater Street,
Suite 906, Ottawa, Ontario K1P 5H3, Canada
(TEL: +1 (613) 233 5623. TELEX: 0534306 CA) (11)

MICHAELI, Prof. I., Department of Polymer Research, Weizmann Institute
of Science, POB 26, Rehovot 76100, Israel
(TEL: +972 (54) 83362. TELEX: 361900 WIX IL) (16)

MICHALSKI, Prof. J., Centrum Badań Molekularnych i Makromolekularnych,
Polska Akademii Nauk, Boczna 5, PL-90 362 Łódź, Poland
(TEL: +48 (4) 41600. TELEX: 886742 PAN PL) (10)

MICHEL, Dr. A., Laboratoire de Matériaux Organiques du Centre National
de la Recherche Scientifique, BP 24, F-69390 Vernaison, France
(TEL: +33 (7) 802 13 44. TELEX: 900225 CNRS VILRB F) (25)

MICHL, Prof. J., Department of Chemistry, University of Utah, Salt Lake City, Utah 84112, USA
(TEL: +1 (801) 581 8549) (22)

MIHAILOVIĆ, Prof. M. Lj., Hemijski Institut, Prirodno-Matematički Fakultet, POB 550, Studentski trg 16, YU-11001 Beograd, Yugoslavia
(TEL: +38 (11) 635104) (22)

MIJS, Dr. W. J., AKZO Research, Postbus 76, 6800 AB Arnhem, Netherlands
(TEL: +31 (85) 663509. TELEX: 45204 NL) (10)

MIKAČ-DEVIĆ, Prof. MIRNA, Department of Clinical Chemistry, University Hospital "Dr. Mladen Stojanović", Vinogradska Cesta 29, YU-41000 Zagreb, Yugoslavia
(TEL: +38 (41) 572730) (35, 36, 37, 38)

MIKI, Dr. T., Production Division, Pharmaceutical Affairs, Takeda Chemical Industries Ltd., 27 Doshomachi 2-chome, Higashi-ku, Osaka, Japan
(TELEX: 63404 TAKEDA J) (12)

MILBURN, Mr. A., Alliance House, 29-30 High Holborn, London WC1V 6BA, UK
(TEL: +44 (1) 404 4083. TELEX: 888941 LCCI G, ATTN. IAWPR) (6)

MILLAN, Dr., Instituto de Plasticos y Caucho, Consejo Superior de Investigaciones Cientificas, Juan de la Cierva 3, Madrid 6, Spain
(TEL: +34 (1) 2622900. TELEX: 42182 CSIC E) (25)

MILLER, Dr. B., 1D-350, Bell Laboratories, 600 Mountain Avenue, Murray Hill, New Jersey 07974, USA
(TEL: +1 (201) 582 6610. TELEX: 13-8650) (16)

MILLS, Prof. I. M., Department of Chemistry, University of Reading, Whiteknights, Reading RG6 2AD, UK
(TEL: +44 (734) 875123. TELEX: 847813 RULIB G) (15)

MIODUSKI, Dr. T., Department of Radiochemistry, Institute of Nuclear Research, PL-03 195 Warsaw, Poland (31)

MITA, Prof. I., Institute of Interdisciplinary Research, Faculty of Engineering, University of Tokyo, 6 Komaba-4, Meguro-ku, Tokyo, 153 Japan
(TEL: +81 (3) 485 3111) (23)

MIYAMOTO, Prof. H., Department of Chemistry, Niigata University, Ikarashi-2, Niigata, 950-21 Japan (31)

MIYAMOTO, Dr. J., Research Department, Pesticide Division, Sumitomo Chemical Co. Ltd., 4-2-1 Takasukasa, Takarazuka, Hyogo, 665 Japan
(TEL: +81 (797) 73 0221. TELEX: 6323 SUMIKA J) (34)

MIZUIKE, Prof. A., Faculty of Engineering, Nagoya University, Chikusa-ku, Nagoya, 464 Japan
(TEL: +81 (52) 781 5111) (27)

MODENA, Prof. G., Istituto di Chimica Organica dell'Università, Via Marzolo 1, I-35100 Padova, Italy
(TEL: +39 (49) 656678. TELEX: 430176 UNIPADUA I) (10, 21)

MØLLER, Mr. A. T., Aarhus Oliefabrik A/S, POB 50, M. P. Bruunsgade 27, DK-8100 Aarhus C, Denmark
(TEL: +45 (6) 12 60 00. TELEX: 64341/2 DK) (33)

MOLIN, Prof. Yu. N., Institute of Chemical Kinetics, Siberian Branch of Academy of Sciences of USSR, SU-630090 Novosibirsk 90, USSR
(TEL: +7 (3832) 654 150) (17)

MONIEN, Prof. H., Universität-Gesamthochschule Siegen, Fachbereich 8, Chemie/Biologie, Analytische Chemie, Postfach 210209, Adolf-Reichwein-Strasse 2, D-5900 Siegen 21, Federal Republic of Germany
(TEL: +49 (271) 740 4248) (29)

MONTALBETTI, Prof. N., Via Keplero 10, I-20124 Milano MI, Italy
(TEL: +39 (2) 6083042. TELEX: 313627 GEXPO I) (13, 35, 36)

MOORE, Dr. D. R., Petrochemicals and Plastics Division, Imperial Chemical Industries PLC, Wilton Centre, F315, POB 90, Wilton, Middlesborough, Cleveland TS6 8JE, UK
(TEL: +44 (642) 455522. TELEX: 587461 G) (24)

MORCILLO RUBIO, Prof. J., Facultad de Ciencias, Universidad de Madrid, Madrid, Spain (12)

MORGAN, Dr. L. G., Medical Department, International Nickel Co. Europe Ltd., Clydach, Swansea SA6 6PJ, UK
(TEL: +44 (792) 843340) (37)

MORITA, Dr. S., Department of Electrical Engineering, Meijo University, Tenpaku-ku, Nagoya, 468 Japan
(TEL: +81 (52) 832 1151) (15)

MORMONT, Ir. R., Laboratoire Central, Solvay & Cie SA, 310 Rue de Ransbeek, B-1120 Bruxelles, Belgium
(TEL: +32 (2) 267 3830. TELEX: 26619 SOLVAY NOH BT B) (10)

MORRIS, Dr. E. G., Section Engineering, Physics and Physical Testing, Great Lakes Research Corpn., POB 1031, Elizabethton, Tennessee 37643, USA
(TEL: +1 (615) 543 3111) (20)

MORRISON, Prof. G. H., Baker Laboratory, Department of Chemistry, Cornell University, Ithaca, New York 14853, USA
(TEL: +1 (607) 256 3614) (28)

MOSS, Dr. G. P., Department of Chemistry, Queen Mary College, Mile End Road, London E1 4NS, UK
(TEL: +44 (1) 980 4811. TELEX: 893750 G) (13, 21, 22)

MROZOWSKI, Prof. S., Department of Physics and Astronomy, Ball State University, Muncie, Indiana 47306, USA
(TEL: +1 (317) 289 1241) (20)

MUKERJEE, Dr. S. K., Division of Agricultural Chemicals, Indian Agricultural Research Institute, New Delhi-110012, India (34)

MÜLLER, Prof. P., Département de Chimie Organique, Université de Genève, 30 quai Ernest-Ansermet, CH-1211 Genève 4, Switzerland
(TEL: +41 (22) 21 93 55) (21, 22)

MULLIKEN, Prof. R. S., Department of Chemistry, University of Chicago, Chicago, Illinois 60637, USA (17)

MUNSCH, Dr. G. B., European Science Foundation, 1 quai Lezay-Marnésia, F-67000 Strasbourg, France
(TEL: +33 (88) 35 30 63. TELEX: 890440 F) (11)

MUNTAU, Mr. H., Commission des Communautés Européennes - EURATOM, I-21020 Ispra (Varese), Italy
(TEL: +39 (332) 780131. TELEX: 380042 EUR I) (37)

MURPHY, Mr. T. J., A347 Chemistry Building, National Bureau of Standards, US Department of Commerce, Washington, DC 20234, USA
(TEL: +1 (301) 921 3674. TELEX: 898493 GARG) (19)

MUSHAK, Dr. P., Department of Pathology, School of Medicine, University of North Carolina, Chapel Hill, North Carolina 27514, USA
(TEL: +1 (919) 966 1395) (37)

MUSSINI, Prof. T., Dipartimento di Chimica Fisica ed Elettrochimica, Università di Milano, Via Golgi 19, I-20133 Milano, Italy
(TEL: +39 (2) 235446) (29)

MYASOEDOV, Prof. B. F., Radiochemical Laboratory, V I Vernadskii Institute of Geochemistry and Analytical Chemistry, Academy of Sciences of USSR, Kosygin Street 19, SU-117975, SZD-1, Moscow, V-334, USSR (27)

MYERS, Prof. R. R., Kent State University, Kent, Ohio 44242, USA (25)

NAGAKURA, Prof. S., Institute for Molecular Science, Myodaiji, Okazaki, Aichi-ken, 444 Japan
(TEL: +81 (564) 52 9770. TELEX: 4537-475 KOKKEN J) (10)

NAGASHIMA, Prof. A., Department of Mechanical Engineering, Faculty of Science and Technology, Keio University, 3-14-1 Hiyoshi, 3-chome, Kohoku-ku, Yokohama, 223 Japan
(TEL: +81 (44) 63 1141. TELEX: 2322198 KOMLIC J) (16)

NAGY, Prof. L. G., Alkalmazott Kémia Tanszék, Budapesti Müszaki Egyetem, Egry József utca 20–22, H-1521 Budapest, Hungary
(TEL: +36 (1) 452 173. TELEX: 225931 MÜEGY H) (18)

NAGY, Dr. T., Magyar Tudományos Akadémia, Központi Kémiai Kutató Intézete, POB 17, Pusztaszeri utca 59-67, H-1525 Budapest, Hungary
(TEL: +36 (1) 353 735. TELEX: 226686 KEMIA H) (25)

NAGYPÁL, Dr. I., Szervetlen és Analitikai Kémiai Tanszék, Kossuth Lajos Tudományegyetem, POB 7, Egyetem tér, H-4010 Debrecen X, Hungary
(TEL: +36 (52) 161 666. TELEX: 72200 UNIVK H) (29, 30)

NAKAJIMA, Prof. A., Department of Polymer Chemistry, Faculty of Engineering, Kyoto University, Yoshida-machi, Sakyo-ku, Kyoto, 606 Japan
(TEL: +81 (75) 751 2111. TELEX: 5422693 LIBKYU J) (23, 25)

NAMBARA, Prof. T., Pharmaceutical Institute, Tohoku University, Aza Aoba, Aramaki, Sendai, 980 Japan
(TEL: +81 (222) 22 1800) (36)

NANCOLLAS, Prof. G. H., Department of Chemistry, Acheson Hall, State University of New York at Buffalo, Main Street Campus, Buffalo, New York 14214, USA
(TEL: +1 (716) 831 3264) (27, 29, 30)

NAUDET, Prof. M., Laboratoire National des Matières Grasses, Faculté des Sciences, Université d'Aix-Marseille, place Victor Hugo, F-13331 Marseille Cédex 3, France
(TEL: +33 (91) 62 45 39) (32, 33)

NAVRATIL, Dr. J. D., Rockwell International, Rocky Flats Plant, POB 464, Golden, Colorado 80401, USA
(TEL: +1 (303) 497 2906) (30)

NEELY, Dr. W. B., Agricultural Products Department, Building 9001, Dow Chemical Co., Midland, Michigan 48640, USA
(TEL: +1 (517) 636 0405) (33)

NEFEDOV, Prof. O. M., N D Zelinsky Institute of Organic Chemistry, Academy of Sciences of USSR, Leninskii Prospekt 47, SU-117913 Moscow, USSR
(TEL: +7 (095) 137 7390) (21)

NEUSE, Prof. E. W., Department of Chemistry, University of the Witwatersrand, 1 Jan Smuts Avenue, Johannesburg 2001, Republic of South Africa
(TEL: +27 (11) 7162311. TELEX: 4-22460 SA) (23)

NEWBOLD, Prof. B. T., Centre Universitaire de Moncton, Université de Moncton, Moncton, Nouveau-Brunswick E1A 3E9, Canada
(TEL: +1 (506) 858 4105. TELEX: 014 353 CA) (11)

NG, Prof. S., Jabatan Kimia, Universiti Malaya, Lembah Pantai, Kuala Lumpur 22-11, Malaysia
(TEL: +60 (3) 555466. TELEX: 37453 MA) (17)

NIEBOER, Dr. E., Department of Biochemistry, Health Sciences Centre, McMaster University, 1200 Main Street West, Hamilton, Ontario L8N 3Z5, Canada
(TEL: +1 (416) 525 9140. TELEX: 0618347 CA) (36, 37, 38)

NIETO DE CASTRO, Prof. C. A., Centro de Química Estrutural, Complexo Interdisciplinar, Instituto Superior Técnico, P-1096 Lisboa-Codex, Portugal
(TEL: +351 (19) 553947) (16)

NIKI, Dr. K., Department of Electrochemistry, Faculty of Engineering, Yokohama National University, 156 Tokiwa-dai, Hodogaya-ku, Yokohama, 240 Japan
(TEL: +81 (45) 335 1451. TELEX: 3822-614 YNUENG J) (16)

NIXON, Dr. J. F., School of Chemistry and Molecular Sciences, University of Sussex, Falmer, Brighton BN1 9QJ, UK
(TEL: +44 (273) 606755) (19)

NOGAR, Dr. N. S., CHM.-2, MS-G 738, Los Alamos National Laboratory, University of California, POB 1663, Los Alamos, New Mexico 87544, USA
(TEL: +1 (505) 667 9305/4686. TELEX: 660495) (28)

NÓGRÁDI, Prof. M., Institute of Organic Chemistry, Technical University, Gellért tér 4, H-1111 Budapest, Hungary (21)

NOMIYAMA, Dr. H., Department of Environmental Health, Jichi Medical School, Minamikawachi-machi, Tochigi-ken, 329-04 Japan
(TEL: +81 (285) 44 2111) (37)

NOMOTO, Dr. S., Division of Clinical Chemistry, School of Medical Technology, Shinshu University, 1308 Asama, Matsumoto, 390-03 Japan
(TEL: +81 (263) 35 4600) (35, 36, 37)

NORSETH, Dr. T., Institute of Occupational Health, Gydas Vei 8, POB 8149 Dep., Oslo 1, Norway
(TEL: +47 (2) 466850) (37)

NÜRNBERG, Prof. H. W., Institut für Chemie der Kernforschungsanlage Jülich GmbH, Institut 4: Angewandte Physikalische Chemie, Postfach 1913, D-5170 Jülich 1, Federal Republic of Germany
(TEL: +49 (2461) 615768/613177. TELEX: 833 556 KFA D) (29)

O'DRISCOLL, Prof. K. F., Department of Chemical Engineering, Faculty of Engineering, University of Waterloo, Waterloo, Ontario N2L 3G1, Canada
(TEL: +1 (519) 885 1211. TELEX: 06955259 CA) (26)

OESTERLE, Dr. K. M., Goldbacherstrasse 88, CH-8700 Küsnacht, Switzerland (25)

OGAREV, Dr. V. A., Institute of Physical Chemistry, Academy of Sciences of USSR, Leninskii Prospekt 31, SU-117312 Moscow, B-312, USSR (25)

O'HARE, Prof. P. A. G., Chemical Engineering Division, Argonne National Laboratory, US Department of Energy, 9700 South Cass Avenue, Argonne, Illinois 60439, USA
(TEL: +1 (312) 972 4535. TELEX: 910 258 3285) (16)

OHLSON, Dr. R., AB Karlshamns Oljefabriker, S-292 00 Karlshamn, Sweden
(TEL: +46 (454) 14300. TELEX: 4510 AKOFAB S) (33)

OHSE, Dr. R. W., Kommission der Europäischen Gemeinschaften, Gemeinsame Forschungsstelle, Europäisches Institut für Transurane, Forschungsanstalt Karlsruhe, Postfach 2266, D-7500 Karlsruhe, Federal Republic of Germany
(TEL: +49 (721) 84384. TELEX: 07825483 EU D) (20)

OHTAKI, Prof. H., Department of Electronic Chemistry, Tokyo Institute of Technology at Nagatsuta, Nagatsuta-cho, Midori-ku, Yokohama, 227 Japan
(TEL: +81 (45) 922 1111. TELEX: 2466360 TITECH J) (29, 30)

ŌKI, Prof. M., Department of Chemistry, Faculty of Science, University of Tokyo, 3-1 Hongo 7-chome, Bunkyo-ku, Tokyo, 113 Japan
(TEL: +81 (3) 812 2111. TELEX: 33659 UTYOSCI J) (21)

OKUDA, Prof. K., Department of Laboratory Medicine and Central Clinical Laboratory, Osaka City University Medical School, Abeno, Osaka, 545 Japan
(TEL: +81 (6) 633 1221) (14, 35)

OLAFSSON, Dr. J., Falconbridge Nikkelverk Aktieselskap, POB 457, N-4601 Kristiansand S, Norway
(TEL: +47 (42) 25500) (37)

OLAJ, Prof. O., Institut für Physikalische Chemie der Universität Wien, Währingerstrasse 42, A-1090 Wien, Austria
(TEL: +43 (222) 343616) (23)

OLDIGES, Dr. H., Fraunhofer-Institut für Toxikologie und Aerosolforschung, Institutsteil Grafschaft, D-5948 Schmallenberg, Federal Republic of Germany (TEL: +49 (2972) 494-96. TELEX: 841520 D) (37)

OLOFSSON, Dr. GERD, Chemical Center, Thermochemistry, University of Lund, Box 740, S-220 07 Lund 7, Sweden (TEL: +46 (46) 108185. TELEX: S-33533 LUNIVER) (16)

O'NEILL, Dr. L. A., Paint Research Association, Waldegrave Road, Teddington, Middlesex TW11 8LD, UK (TEL: +44 (1) 977 4427) (25)

ONG, Prof. S. H., Chemistry and Technology Division, PORIM, 18th Floor, Angkasa Raya, Jalan Ampang, Kuala Lumpur, Malaysia (33)

ONKELINX, Dr. C., Rue des Floralies 85/20, B-1200 Bruxelles, Belgium (TEL: +32 (2) 46 68 50) (36)

ONO, Dr. T., Research and Development Division, Toray Industries Inc., 2-2 Nihonbashi-muromachi, Chuo-ku, Tokyo, 103 Japan (TEL: +81 (3) 245 5306. TELEX: 22623 J) (24)

O'REILLY, Mr. J. M., Webster Research Center, Xerox Corpn., 800 Phillips Road, Webster, New York 14580, USA (26)

OREKHOVICH, Prof. V. N., Institute of Biological and Medical Chemistry, Academy of Medical Sciences of USSR, Pogodynskaya Street 10, SU-119832 Moscow, G-121, USSR (35)

DEN OS, Dr. D. P., Bureau van de Rijksuniversiteit te Leiden, Postbus 9500, 2300 RA Leiden, Netherlands (TEL: +31 (71) 148333. TELEX: 39427 BURUL NL) (11)

OSMAN, Prof. A. M., Suez Canal University, Ismailia, Arab Republic of Egypt (TEL: Cairo 600844. TELEX: 93648 NCC UN) (22)

OSTERMANN, Dr. G., Union Deutsche Lebensmittelwerke GmbH, Zweigniederlassung Hamburg-Bahrenfeld, Postfach 501020, Friedensallee 333, D-2000 Hamburg 50, Federal Republic of Germany (TEL: +49 (40) 8821515. TELEX: 0213879 D) (33)

OSTERYOUNG, Prof. JANET G., Department of Chemistry, Faculty of Natural Sciences and Mathematics, Acheson Hall, State University of New York at Buffalo, Buffalo, New York 14214, USA (TEL: +1 (716) 831 3014) (29)

O'SULLIVAN, Dr. D. A., Chemical and Engineering News (Europe), 3 Sheridan Road, London SW19 3HW, UK (TEL: +44 (1) 540 0414) (11)

OTTAWAY, Prof. J. M., Department of Pure and Applied Chemistry, Thomas Graham Building, University of Strathclyde, 295 Cathedral Street, Glasgow G1 1XL, UK (TEL: +44 (41) 552 4400) (27, 37)

OTTO, Dr. S., Landwirtschaftliche Versuchsstation, BASF Aktiengesellschaft, Postfach 220, D-6703 Limburgerhof, Federal Republic of Germany (TEL: +49 (6236) 68201. TELEX: 464600 BASF D) (33)

OVERBERGER, Prof. C. G., 4080 Administration Building, University of Michigan, Ann Arbor, Michigan 48109, USA (TEL: +1 (313) 764 1185. TELEX: 8102236056 U OF M AA) (10)

ØYE, Prof. H. A., Institutt for Uorganisk Kjemi, Universitet i Trondheim-Norges Tekniske Høgskole, N-7034 Trondheim-NTH, Norway (TEL: +47 (75) 94016. TELEX: 55637 NTHAD N, ATTN. UORGANISK) (16)

PADOVA, Dr. I., Israel Atomic Energy Commission, Soreq Nuclear Research Center, Javne 70600, Israel (30)

PAGANI, Prof. D., Istituto di Chimica, Politecnico di Milano, Piazza Leonardo da Vinci 32, I-20133 Milano, Italy (TELEX: 333467 POLIMI I) (25)

PAJARES, Dr. J. A., Instituto de Catalisis y Petroleoquimica, Consejo Superior de Investigaciones Científicas, Serrano 119, Madrid 6, Spain (TEL: +34 (1) 2619400. TELEX: 42182 CSIC E) (18)

PALMA, Dr., Montedison CER, Via Motta, Porto Marghera (Venezia), Italy (25)

PANAYE, Prof. ANNICK, Institut de Topologie et de Dynamique des Systèmes, Centre National de la Recherche Scientifique, Université de Paris VII, 1 rue Guy de la Brosse, F-75005 Paris, France (TEL: +33 (1) 336 25 25) (22)

PANAYOTOV, Prof. I. M., Central Laboratory for Polymers, Bulgarian Academy of Sciences, Sofia 1113, Bulgaria (23)

PANICO, Prof. R., Direction des Etudes, Ecole Supérieure de Physique et de Chimie Industrielles, Université de Paris VI, 10 rue Vauquelin, F-75231 Paris Cédex 05, France (TEL: +33 (1) 337 77 00) (21)

PAOLETTI, Prof. P., Dipartimento di Chimica, Università di Firenze, Via J. Nardi 39, I-50132 Firenze, Italy (TEL: +39 (55) 242922/240842/3/5) (29)

PAOLONI, Prof. L., Gruppo Chimica Teorica, Università di Palermo, Via Archirafi 20, I-90123 Palermo, Italy (TEL: +39 (91) 281143. TELEX: 910170 UNIVPA I) (11, 12)

PAPISOV, Prof. I. M., Chemistry Department, Moscow Automobile and Road Construction Institute, Leningradsky Prospekt 64, Moscow, A-319, USSR (TEL: +7 (095) 155 0321/0826) (23)

PAREDES, Dr. E., Unidad de Petroquímica, INTEVEP, Apartado 76343, Caracas 1070-A, Venezuela (TELEX: 21672 INTVP VE) (23)

PARISER, Dr. R., Polymer Science, Central Research and Development Department, E. I. du Pont de Nemours & Co. Inc., Experimental Station, 1007 Market Street, Wilmington, Delaware 19898, USA (TEL: +1 (302) 772 1800. TELEX: 835420 DUPONT) (10)

PARISI, Prof. F., Società Chimica Italiana - Sezione Lombarda, Piazzale Rodolfo Morandi 2, I-20121 Milano, Italy (33)

PARK, Dr., University of Wales Institute of Science and Technology, King Edward VII Avenue, Cardiff CF1 3NU, UK (TEL: +44 (222) 42522) (25)

PARK, Dr. D. L., HFF 454, Organic and Biological Chemistry Branch, Division of Chemistry and Physics, Bureau of Foods, Food and Drug Administration, US Department of Health and Human Services, 200 C Street SW, Washington, DC 20204, USA (TEL: +1 (202) 245 1453. TELEX: 898488 PHS PKLN ROVE) (32)

PARSONS, Dr. R., Department of Physical Chemistry, School of Chemistry, University of Bristol, Cantock's Close, Bristol BS8 1TS, UK (TEL: +44 (272) 24161. TELEX: 444174 BUPHYS G) (15, 16)

PARUTA, Dr. A. N., College of Pharmacy, University of Rhode Island, Room 147, Fogarty Hall, Kingston, Rhode Island 02882, USA (30)

PAUL, Dr. R. L., Electrochemistry, Mineral and Process Chemistry Division, MINTEK, Private Bag X3015, Randburg 2125, Republic of South Africa (TEL: +27 (11) 7933511. TELEX: 4-24867 SA) (16)

PAULIK, Dr. J., Institut für Allgemeine und Analytische Chemie, Technische Universität Budapest, Gellért tér 4, H-1111 Budapest, Hungary (26)

PEARCE, Prof. E. M., Polymer Research Institute, Polytechnic Institute of New York, 333 Jay Street, Brooklyn, New York 11201, USA (TEL: +1 (212) 643 5470) (23)

PECHEN DE D'ANGELO, Dr. ANA MARÍA, Mitre 677, Neuquén, Prov. Neuquén, Argentina (34)

PEISER, Dr. H. S., 638 Blossom Drive, Rockville, Maryland 20850, USA (TEL: +1 (301) 762 6860. TELEX: 904059 WSH) (19)

PELLERIN, Prof. F., Faculté de Pharmacie de l'Université de Paris-Sud, Hôpital Corentin-Celton, 37 boulevard Gambetta, F-92130 Issy les Moulineaux, France (TEL: +33 (1) 554 95 33. TELEX: 613145 SOBIO F) (10, 11, 27)

PENASSE, Dr. L., Direction des Recherches Biologiques et Biochimiques Avancées, Division Scientifique Roussel Uclaf SA, 111 route de Noisy, F-93230 Romainville, France (TEL: +33 (1) 843 93 10. TELEX: 680501 ROUSSEL RMVI F) (32)

PENTON, Dr. J. R., Dorfstrasse 49, CH-8037 Zürich, Switzerland (21)

PEPPER, Prof. D. C., Chemical Laboratory, Trinity College, University of Dublin, Dublin 2, Ireland (TEL: +0044 (0001) 772941. TELEX: 25442 TCD EI) (23)

PEREIRA, Dr. MARIANA P. B. A., Av. Combatentes da Grande Guerra - 15-1°D, P-2700 Amadora, Portugal (TEL: +351 (19) 42542) (12)

PERIN, Miss REYHAN, Unilever-Iş Türk Ltd. Şti., Cumhuriyet Caddesi, Kahan Apt. 16, Harbiye-Istanbul, Turkey (TEL: +90 (11) 406233/466156. TELEX: 22467 TR) (33)

PERONE, Prof. S. P., Lawrence Livermore National Laboratory, L-326, Chemistry and Materials Science Department, University of California, POB 808, Livermore, California 94550, USA (TEL: +1 (415) 422 6322) (28)

PERRIN, Dr. D. D., Medical Chemistry Group, John Curtin School of Medical Research, Australian National University, PO Box, Canberra, ACT 2601, Australia (TEL: +61 (62) 49 3526) (29, 30)

PERSON, Prof. W. B., Department of Chemistry, University of Florida, Gainesville, Florida 32611, USA (TEL: +1 (904) 392 0528. TELEX: 8108256320 TWX) (17)

PESCH, Prof. H.-J., Pathologisches Institut der Universität Erlangen-Nürnberg, Krankenhausstrasse 8-10, D-8520 Erlangen, Federal Republic of Germany (TEL: +49 (9131) 852289) (37)

PETLEY, Dr. B. W., National Physical Laboratory, Teddington, Middlesex TW11 OLW, UK (12)

PETRII, Prof. O. A., Faculty of Chemistry, Lomonosov Moscow State University, Leninskie Gory, SU-117234 Moscow, V-234, USSR (TEL: +7 (095) 139 2570/5501) (16)

PETTIT, Dr. L. D., School of Chemistry, University of Leeds, Leeds LS2 9JT, UK (TEL: +44 (532) 431 751) (29, 30)

PETZOW, Prof. G., Max-Planck-Institut für Metallforschung, Heisenbergstrasse 5, D-7000 Stuttgart 80, Federal Republic of Germany (TEL: +49 (711) 681476. TELEX: 07 23742 D) (20)

PFEIL, Prof. W., Forschungszentrum für Molekularbiologie und Medizin, Zentralinstitut für Molekularbiologie, Akademie der Wissenschaften der DDR, Robert-Rössle-Strasse 10, DDR-1115 Berlin, German Democratic Republic (TELEX: 0112625 DD) (13)

PFENDER, Prof. E., Department of Mechanical Engineering, University of Minnesota, 111 Church Street SE, Minneapolis, Minnesota 55455, USA (15)

PHAM, Dr. Q. T., Laboratoire Central de Microanalyse du Centre National de la Recherche Scientifique, BP 24, F-69390 Vernaison, France (TEL: +33 (7) 802 13 44. TELEX: 900225 CNRS VILRB F) (25)

PHILIPP, Prof. B., Institut für Polymerenchemie "Erich Correns", Akademie der Wissenschaften der DDR, Kantstrasse 55, DDR-1530 Teltow-Seehof, German Democratic Republic
(TEL: 4831. TELEX: 15322 DD) (23, 25)

PICK, Prof. J., Department of Physical Chemistry, Vysoká Škola Chemicko-Technologická v Praze, Suchbátarova 5, CS-166 28 Praha 6-Dejvice, Czechoslovakia
(TELEX: 122744 VSCH C) (16)

PIEKOS, Dr. R., Faculty of Pharmacy, Medical Academy, ul. Marksa 107, PL-80 416 Gdańsk, Poland
(TEL: +48 (98) 418011) (31)

PIERCE, Dr. P. E., PPG Industries Inc., POB 9, Rosanna Drive, Allison Park, Pennsylvania 15101, USA (25)

PIETSCH, Prof. H., Akademie der Wissenschaften der DDR, Haus 10.2, Rudower Chaussee, DDR-1199 Berlin, German Democratic Republic
(TEL: +37 (2) 6702841) (6)

PIKKARAINEN, Dr. J., National Public Health Institute, Mannerheimintie 166, SF-00280 Helsinki 28, Finland (38)

PILJAC, Prof. I., Prehrambeno-biotehnološki fakultet, Sveučilišta u Zagrebu, Pierottijeva 6, YU-41000 Zagreb, Yugoslavia
(TEL: +38 (41) 440722) (29)

PINO, Prof. P., Institut für Polymere, Eidgenössische Technische Hochschule Zürich, Universitätstrasse 6, CH-8092 Zürich, Switzerland
(TEL: +41 (1) 256 30 38. TELEX: 53178 ETHBI CH) (23)

PLAA, Prof. G., Département de Pharmacologie, Faculté de Médecine, Université de Montréal, Case Postale 6128, Succursale A, Montréal, Québec, Canada
(TEL: +1 (514) 343 6300) (37)

PLAMBECK, Dr. J. A., Department of Chemistry, University of Alberta, Edmonton, Alberta T6G 2E1, Canada
(TEL: +1 (403) 432 3253) (16)

PLATÉ, Prof. N. A., Institute of Petrochemical Synthesis, Academy of Sciences of USSR, Leninskii Prospekt 29, SU-117912 Moscow, USSR
(TEL: +7 (095) 139 1189) (23, 24)

PLOCHOCKI, Dr. A., Chemical Engineering Department, Stevens Institute of Technology, Castle Point, Hoboken, New Jersey 07030, USA
(TEL: +1 (201) 420 5773) (24)

PLŠKO, Prof. E., Geologický Ústav Univerzity Komenského, Zadunajská 15, CS-851 01 Bratislava, Czechoslovakia
(TEL: +42 (7) 53858) (28)

POCKLINGTON, Mr. W. D., Laboratory of the Government Chemist, Cornwall House, Stamford Street, London SE1 9NQ, UK
(TEL: +44 (1) 928 7900. TELEX: 881314 DIHQG LAB GOV CHEM G) (33)

POHLAND, Dr. A. E., HFF 452, Organic and Biological Chemistry Branch, Division of Chemistry and Physics, Bureau of Foods, Food and Drug Administration, US Department of Health and Human Services, 200 C Street SW, Washington, DC 20204, USA
(TEL: +1 (202) 245 1036. TELEX: 7108229530) (32)

POKORNÝ, Prof. J., Katedra Chemie a Zkoušení Potravin, Vysoká Škola Chemicko-Technologická v Praze, Suchbátarova 5, CS-166 28 Praha 6-Dejvice, Czechoslovakia
(TEL: +42 (2) 332. TELEX: 122744 VSCH C) (33)

PONNAMPERUMA, Prof. C., Laboratory of Chemical Evolution, Department of Chemistry, University of Maryland, College Park, Maryland 20742, USA
(TEL: +1 (301) 454 5676/5677. TELEX: 908787 OHY UNMD CORK) (13)

PORTER, Dr. C. J., ES 3-404D, Toronto General Hospital, 200 Elizabeth Street, Toronto, Ontario M5G 1L7, Canada
(TEL: +1 (416) 595 3283) (36)

POWELL, Dr. H. K. J., Chemistry Department, University of Canterbury, Private Bag, Christchurch, New Zealand (30)

POWELL, Dr. W. H., Chemical Abstracts Service, POB 3012, 2540 Olentangy River Road, Columbus, Ohio 43210, USA
(TEL: +1 (614) 421 3632. TELEX: 684-2086) (12, 21)

PROTIVA, Prof. M., Výzkumný Ústav pro Farmacii a Biochemii, Kouřimská 17, CS-130 00 Praha 3-Vinohrady, Czechoslovakia
(TEL: +42 (2) 734941. TELEX: 122554 VUFB C) (21)

PSZONICKI, Dr. L., International Atomic Energy Agency, Laboratory Seibersdorf, POB 100, Wagramerstrasse 5, A-1400 Wien, Austria
(TEL: +43 (2254) 2251. TELEX: 1-12645 A) (28)

PUJADE-RENAUD, Dr. J. M., Département de Analyse, Minemet Recherche (Group IMETAL), BP 106 ZI, 1 avenue Albert Einstein, F-78190 Trappes Cédex, France
(TEL: +33 (3) 050 61 88. TELEX: MITEC 696745 F) (37)

PUNGOR, Prof. E., Altalános és Analitikai Kémiai Tanszék, Budapesti Müszaki Egyetem, Gellért tér 4, H-1521 Budapest, Hungary
(TELEX: 225931 MÜEGY H) (27, 29)

PURDELA, Dr. D., Centrul de Chimie - Timişoara, Bul. Mihai Viteazul Nr. 24, 1900 Timişoara, Romania (19)

PURDY, Prof. W. C., Department of Chemistry, McGill University, 801 Sherbrooke Street West, Montreal, Québec H3A 2K6, Canada
(TEL: +1 (514) 392 4437) (11, 29, 36)

QIAN, Prof. R., Institute of Chemistry, Chinese Academy of Sciences, Chung Kuan Tsun, Beijing, China
(TEL: 284472) (24)

QUINN, Dr. T. J., Bureau International des Poids et Mesures, pavillon de Breteuil, 12 bis Grande Rue, F-92310 Sèvres, France
(TEL: +33 (1) 534 00 51) (12)

QUIRK, Prof. R. P., Institute of Polymer Science, University of Akron, Akron, Ohio 44325, USA
(TEL: +1 (216) 375 7500) (24)

RAASCHOU NIELSEN, Mr. H. K., Scandinavian Paint and Printing Ink Research Laboratory, Agern Alle 3, DK-2970 Hørsholm, Denmark (25)

RABOCKAY, Prof. T., Instituto de Química, Universidade de São Paulo, Cidade Universitária, 01000 São Paulo - SP, Brazil (16)

RÅNBY, Prof. B., Department of Polymer Technology, Royal Institute of Technology, S-100 44 Stockholm, Sweden
(TEL: +46 (8) 7878265. TELEX: 10389 KTHB STOCKHOLM S) (23)

RANCKE-MADSEN, Prof. E., Kemisk Institut, Danmarks Laererhøjskole, Emdrupvej 115 B, DK-2400 København NV, Denmark
(TEL: +45 (1) 69 6633) (11)

RAND, Dr. M. H., Materials Development Division, Building 552, Atomic Energy Research Establishment, Harwell, Didcot, Oxfordshire OX11 ORA, UK
(TEL: +44 (235) 24141. TELEX: 83135 G) (20)

RANGARAJAN, Prof. S. K., Department of Inorganic and Physical Chemistry, Indian Institute of Science, Malleswaram, Bangalore-560012, India
(TEL: 34411. TELEX: 0845-8349 IISC IN) (16)

RAO, Prof. C. N. R., Solid State and Structural Chemistry Unit, Indian Institute of Science, Malleswaram, Bangalore-560012, India
(TEL: 34411. TELEX: 0845-8349 IISC IN) (v, 10, 13)

RASBERRY, Mr. S. D., Office of Standard Reference Data, National Bureau of Standards, US Department of Commerce, Washington, DC 20234, USA
(TEL: +1 (301) 921 2408. TELEX: 898493 GARG) (15)

RATAJCZAK, Prof. H., Instytut Chemii, Uniwersytet Wrocławski, ul. Joliot-Curie 14, PL-50 383 Wrocław, Poland
(TEL: +48 (7) 222348/229281. TELEX: 0712791 PL) (17)

RÄTZSCH, Prof. MARGIT T., Technische Hochschule "Carl Schorlemmer" Leuna-Merseburg, Sektion Chemie, Geusaer Strasse, DDR-4200 Merseburg 6, German Democratic Republic
(TEL: +37 (442) 462372. TELEX: 471320 THLM DD) (16)

REEDIJK, Prof. J., Gorlaeus Laboratories, Department of Chemistry, State University Leiden, PO Box 9502, 2300 RA Leiden, Netherlands
(TEL: +31 (71) 148333. TELEX: 39427 BURUL NL) (12, 13, 19)

REES, Dr. C. E., Department of Geology, McMaster University, 1280 Main Street West, Hamilton, Ontario L8S 4K1, Canada
(TEL: +1 (416) 525 9140. TELEX: 061-8347 CA) (19)

REGOSZ, Dr. A., Faculty of Pharmacy, Medical Academy, ul. Marksa 107, PL-80 416 Gdańsk, Poland (31)

REILLY, Prof. P. M., Department of Chemical Engineering, Faculty of Engineering, University of Waterloo, Waterloo, Ontario N2L 3G1, Canada
(TEL: +1 (519) 885 1211. TELEX: 06955259 CA) (26)

REINISCH, Prof. G., Institut für Polymerenchemie "Erich Correns", Akademie der Wissenschaften der DDR, Kantstrasse 55, DDR-1530 Teltow-Seehof, German Democratic Republic
(TEL: 4831. TELEX: 15322 DD) (23)

REITH, Dr. A. K., Norsk Hydros Institute for Cancer Research, Norwegian Radium Hospital, Montebello, Oslo 3, Norway
(TEL: +47 (2) 554080) (37)

REMPP, Prof. P., Centre de Recherches sur les Macromolécules du Centre National de la Recherche Scientifique, 6 rue Boussingault, F-67083 Strasbourg Cédex, France
(TEL: +33 (88) 61 19 19) (26)

RETAMAR, Prof. J. A., Instituto de Investigaciones de Productos Naturales de Análisis y Síntesis Orgánica, CONICET-UNL-FIQ, Santiago del Estero 2829, 3000 Santa Fe, Argentina
(TEL: +54 (42) 40758. TELEX: 48 153 INLIT AR) (21)

RETTING, Dr. W., BASF Aktiengesellschaft, D-ZKM-G201, D-6700 Ludwigshafen/Rhein, Federal Republic of Germany
(TEL: +49 (621) 6028797. TELEX: 464811 BASF D) (24)

RHYNER, Dr. P., K25.505, CIBA-GEIGY AG, Klybeckstrasse 141, CH-4002 Basel, Switzerland
(TEL: +41 (61) 36 37 48. TELEX: 62991 CIGY CH) (10)

RIALDI, Prof. G., Istituto di Chimica Industriale, Via A. Pastore 3, Genoa, Italy
(TEL: +39 (10) 518141) (13)

RIBAS, Prof. B., Faculdad de Farmacia, Instituto de Bioquimica, Universidad Complutense de Madrid, Ciudad Universitaria, Madrid 3, Spain
(TEL: +34 (1) 2438649/6262. TELEX: 42182 CSIC E) (35)

RICHARDS, Prof. D. H., Ministry of Defence, Propellants, Explosives and Rocket Motor Establishment, Sewardstone Road, Waltham Abbey, Essex EN9 1BP, UK
(TEL: +44 (992) 713030. TELEX: 267455 G) (23)

RICHARDS, Prof. G. N., Department of Chemistry and Biochemistry, James Cook University of North Queensland, Post Office, James Cook University, Queensland 4811, Australia
(TEL: Douglas 814343. TELEX: AA47009) (6)

RICHARDSON, Dr. J. M., National Physical Laboratory, Department of Industry, Teddington, Middlesex TW11 0LW, UK (26)

RICHER, Prof. J. C., Département de Chimie, Université de Montréal, Case Postale 6210, Succursale A, Montréal, Québec H3C 3V1, Canada
(TEL: +1 (514) 343 6744) (21, 22)

RICHTER, Dr. U. P., Chemische Werke Hüls AG, ZB Forschung und Entwicklung, Postfach 1320, D-4370 Marl-Hüls 1, Federal Republic of Germany (TEL: +49 (2365) 495210. TELEX: 829211-0 CW D) (24)

RIEKERT, Prof. L., Institut für Chemische Verfahrenstechnik, Universität Karlsruhe (TH), Postfach 6380, Kaiserstrasse 12, D-7500 Karlsruhe 1, Federal Republic of Germany (TEL: +49 (721) 608 3947. TELEX: 07826521 UNI D) (18)

RIGAUDY, Prof. J., Laboratoire de Recherches Organiques, Ecole Supérieure de Physique et de Chimie Industrielles, Université de Paris VI, 10 rue Vauquelin, F-75231 Paris Cédex 05, France (TEL: +33 (1) 707 88 36) (21)

RIGG, Mr. J. C., Breehoven 90, 6721 SG Bennekom, Netherlands (TEL: +31 (8389) 9373) (35)

RIGHELATO, Dr. R. C., Tate and Lyle Ltd., Group Research and Development, Philip Lyle Memorial Research Laboratory, University of Reading, POB 68, Whiteknights Park, Reading RG6 2BX, UK (TEL: +44 (734) 861361. TELEX: 847915 TALRES G) (32)

RINAUDO, Prof. MARGUERITE, Centre de Recherches sur les Macromolécules Végétales, Centre National de la Recherche Scientifique 53 X, F-38041 Grenoble Cédex, France (TEL: +33 (76) 54 11 45. TELEX: 320254 CNRS ALPGRENO F) (24)

RING, Prof. W., Chemische Werke Hüls AG, Unternehmensleitung, Postfach 1320, D-4370 Marl-Hüls 1, Federal Republic of Germany (TEL: +49 (2365) 485252. TELEX: 829211-OCW D) (12, 23)

RINGELHANN, Dr. B., Országos Reuma és Fizikoterápiás Intézet, POB 54, Frankel Leó utca 17-19, H-1525 Budapest, Hungary (35)

RINGNES, Mrs. VIVI, School Science Center, Institute of Chemistry, University of Oslo, Box 1033, Blindern, Oslo 3, Norway (TEL: +47 (2) 455579) (12)

RINGPFEIL, Prof. M., Institut für Technische Chemie der Akademie der Wissenschaften der DDR, Permoserstrasse 15, DDR-7050 Leipzig, German Democratic Republic (TEL: +37 (41) 2392225. TELEX: 051536 DD) (32)

RISTIĆ, Prof. M. M., Division of Physics and Technology of Sintering and Sintered Materials, Institute of Technical Sciences, Serbian Academy of Sciences and Arts, POB 745, Knez Mihailova 35/IV, YU-11000 Beograd, Yugoslavia (TEL: +38 (11) 637601) (20)

RIVAIL, Prof. J. L., Laboratoire de Chimie Theorique, Université de Nancy I, BP 239, F-54506 Vandoeuvre les Nancy, France (TEL: +33 (8) 328 93 93) (11)

ROBBINS, Dr. D. A., Department of Environmental Sciences, ASARCO Inc., 3422 South 700 West, Salt Lake City, Utah 84119, USA (TEL: +1 (801) 262 2459) (37)

ROBERTS, Prof. M. W., Department of Physical Chemistry, University College, POB 78, Cardiff CF1 1XL, UK (TEL: +44 (222) 44211. TELEX: 498635 G) (18)

ROBERTS, Dr. T. R., Shell Research PLC, Broad Oak Road, Sittingbourne, Kent ME9 8AG, UK (TEL: +44 (795) 24444) (33)

ROBIN, Prof. J., Bâtiment 401, Laboratoire de Physicochimie Industrielle, Institut National des Sciences Appliquées de Lyon, 20 avenue Albert Einstein, F-69621 Villeurbanne Cédex, France (TEL: +33 (7) 893 81 12. TELEX: 380856 INSALYN F) (27)

ROBINSON, Dr. MARION F., Department of Nutrition, University of Otago, POB 56, Dunedin, New Zealand (TEL: +64 (24) 771640) (38)

ROLZ, Ing. C., c/o ICAITI, POB 1552, Guatemala City, Guatemala (6)

RONCA, Dr. P. V., Chemistry and Insulations Branch, Engineering Laboratory, Canadian General Electric Co. Ltd., 107 Park Street North, Peterborough, Ontario K9J 7B5, Canada
(TEL: +1 (705) 748 7721. TELEX: 06 962826 CANEDISON CA) (26)

RONDININI CAVALLARI, Prof. SANDRA, Dipartimento di Chimica Fisica ed Elettrochimica, Università di Milano, Via Venezian 21, I-20133 Milano, Italy
(TEL: +39 (2) 293516. TELEX: 320484 UNI MI I) (29)

ROSENBLATT, Prof. G. M., Chemistry Division, MS-J 563, Los Alamos National Laboratory, University of California, POB 1663, Los Alamos, New Mexico 87544, USA
(TEL: +1 (505) 667 6250. TELEX: 660495) (20)

ROSENTHAL, Dr. I., Department of Food Technology, Agricultural Research Organization, Ministry of Agriculture, Volcani Center, POB 6, Bet Dagan 50250, Israel
(TEL: +972 (3) 980421/259. TELEX: 341118 BXTV IL, EXT. 5430) (32)

ROSMAN, Dr. K. J. R., School of Physics and Geosciences, Western Australian Institute of Technology, Kent Street, Bentley, Western Australia 6102, Australia
(TEL: +61 (9) 350 7542. TELEX: AA92983) (19)

ROSSET, Prof. R. H., Laboratoire de Chimie Analytique des Processus Industriels, Ecole Supérieure de Physique et de Chimie de Paris, 10 rue de Vauquelin, F-75231 Paris Cédex 05, France
(TEL: +33 (1) 337 77 00/535 00 04) (28)

ROSSI, Prof. R. A., Facultad de Ciencias Quimicas, Universidad Nacional de Cordoba, Cordoba, Argentina
(TEL: +54 (51) 62512. TELEX: 51822 BUCOR AR) (21)

ROSSITER, Dr. B. W., Research Laboratories, B-82, Eastman Kodak Co., 1669 Lake Avenue, Rochester, New York 14650, USA
(TEL: +1 (716) 722 2955/458 2599. TELEX: 978323 KODAKDIST) (10)

ROTH, Prof. E., Division d'Etudes de Separation Isotopique et de Chimie Physique, Commissariat à l'Energie Atomique, Centre d'Etudes Nucléaires de Saclay, BP 2, F-91191 Gif-sur-Yvette, France
(TEL: +33 (6) 908 3405. TELEX: 20671 ENERGAT PARIS F) (17)

ROTH, Dr. M., Laboratoire de Chimie Clinique, Hôpital Universitaire, CH-1211 Genève 4, Switzerland
(TEL: +41 (22) 22 75 51. TELEX: 427588 HCGE CHGL CH) (12, 36)

ROUQUÉROL, Dr. J., Centre de Thermodynamique et de Microcalorimétrie du Centre National de la Recherche Scientifique, 26 rue du 141e RIA, F-13003 Marseille, France
(TEL: +33 (91) 62 78 74) (16, 17)

ROUSSEL, Dr. D., Societé Chimique des Charbonnages, CdF Chimie, Usines du Nord, BP 1, F-62160 Bully les Mines, France
(TEL: +33 (21) 79 92 33. TELEX: 120142 F) (25)

ROY, Mr. B. R., AMAX Center, AMAX Inc., Greenwich, Connecticut 06830, USA
(TEL: +1 (203) 622 3034) (37)

ROZPLOCH, Dr. F., Instytut Fizyki, Uniwersitet Mikolaja Kopernika, ul. Grudsziadzka 5, PL-87 100 Toruń, Poland
(TEL: +48 (56) 25374) (20)

RUBAJLO, Dr. V. L., Institute of Chemical Physics, Academy of Sciences of USSR, Kosygin Street 4, SU-117977, SZD-1, Moscow, V-334, USSR (33)

RUBANYI, Dr. G., Department of Physiology, University of Cincinnati, 231 Bethesda Avenue (#576), Cincinnati, Ohio 45267, USA (37)

RUBEL, Prof. S., Wydział Chemii, Uniwersytet Warszawski, ul. Pasteura 1, PL-02 093 Warszawa, Poland
(TEL: +48 (22) 220211. TELEX: 815439 UW PL) (29, 37)

RUBEŠKA, Dr. I., c/o UNDP, POB 7285 ADC, Pasay City, Metro Manila, Philippines
(TELEX: 72222250, 72222251 RCA PH) (28)

RUSSO, Prof. S., Istituto di Chimica Industriale, Università Genova, Corso Europa 30, I-16132 Genova, Italy
(TEL: +39 (10) 518141/3) (26)

RUTKOWSKI, Prof. A., Instytut Technologii Żywności, Akademia Rolnicza, ul. Grochowska 272, PL-03 849 Warszawa, Poland (32)

RUYTER, Dr. I. E., NIOM-Scandinavian Institute of Dental Materials, Forskningsveien 1, Oslo 3, Norway
(TEL: +47 (2) 695880) (23)

RYAN, Prof. D. E., Trace Analysis Research Centre, Department of Chemistry, Dalhousie University, Life Sciences Building, Halifax, Nova Scotia B3H 4J1, Canada
(TEL: +1 (902) 424 7070. TELEX: 019 21863 CA) (27)

SAARINEN, Prof. H., Department of Chemistry, University of Helsinki, Vuorikatu 20, SF-00100 Helsinki 10, Finland (11)

SABBAH, Dr. R., Centre de Thermodynamique et de Microcalorimétrie du Centre National de la Recherche Scientifique, 26 rue du 141e RIA, F-13003 Marseille, France
(TEL: +33 (91) 62 78 74) (15)

SABBIONI, Dr. E., Chemistry Division, Commission of the European Communities, Joint Research Centre, Ispra Establishment, I-21020 Ispra (Varese), Italy
(TEL: +39 (332) 780131. TELEX: 380042 EUR I) (37)

SAEGUSA, Prof. T., Department of Synthetic Chemistry, Kyoto University, Yoshida-machi, Sakyo-ku, Kyoto, 606 Japan
(TEL: +81 (75) 751 2111. TELEX: 5422693 LIBKYU J) (23)

SAITO, Prof. Kazuo, Department of Chemistry, Faculty of Science, Tohoku University, Aza Aoba, Aramaki, Sendai, 980 Japan
(TEL: +81 (222) 22 1800) (19)

SALBU, Dr. BRIT, Department of Chemistry, University of Oslo, POB 1033, Blindern, Oslo 3, Norway
(TEL: +47 (2) 455482) (27)

SALOMON, Dr. M., US Army Electronics Research and Development Laboratories, Power Sources Laboratory (DELET-PR), Fort Monmouth, New Jersey 07703, USA
(TEL: +1 (201) 544 2832) (30, 31)

SALVATORE, Prof. F., Istituto di Chimica Biologica, II Facoltà di Medicina e Chirurgia, Università degli Studi, Via Sergio Pansini 5, I-80131 Napoli, Italy
(TEL: +39 (81) 253026-7. TELEX: 610076 CNR RM I) (35)

SAMUEL, Prof. D., Weizmann Institute of Science, Rehovot 76100, Israel
(TEL: +972 (54) 83117. TELEX: 361900 WIX IL) (11)

SAMUEL, Dr. E., Laboratoire de Chimie Organique Industrielle, Ecole Nationale Supérieure de Chimie, 11 rue Pierre et Marie Curie, F-75005 Paris Cédex 05, France
(TEL: +33 (1) 326 95 55) (19)

SAMUELSSON, Dr. K., Acta Chemica Scandinavica, Wallingatan 26 A, S-111 24 Stockholm, Sweden
(TEL: +46 (8) 207604) (19)

SANDORFY, Prof. C., Département de Chimie, Faculté des Arts et des Sciences, Université de Montréal, Case Postale 6210, Succursale A, Montréal, Québec H3C 3V1, Canada
(TEL: +1 (514) 343 6805) (22)

SANKAR DAS, Dr. M., Analytical Division, Bhabha Atomic Research Centre, Trombay, Bombay-400085, India
(TEL: +91 (22) 525778. TELEX: 0112817 IN) (27)

SARGESON, Prof. A. M., Research School of Chemistry, Australian National University, GPO Box 4, Canberra, ACT 2601, Australia
(TEL: +61 (62) 49 3718. TELEX: AA62172) (19)

SARKAR, Dr. B., Research Institute, Hospital for Sick Children, 555 University Avenue, Toronto, Ontario M5G 1X8, Canada (37)

SAVORY, Mrs. GERALDINE, Clinical Laboratories, Department of Pathology, University of Virginia Medical Center, POB 168, Charlottesville, Virginia 22908, USA
(TEL: +1 (804) 924 2553/5682) (37)

SAVORY, Prof. J., Clinical Laboratories, Department of Pathology, University of Virginia Medical Center, POB 168, Charlottesville, Virginia 22908, USA
(TEL: +1 (804) 924 2553/5682) (36, 37, 38)

SAVVIN, Prof. S. B., V I Vernadskii Institute of Geochemistry and Analytical Chemistry, Academy of Sciences of USSR, Kosygin Street 19, SU-117975, SZD-1, Moscow, V-334, USSR (27)

SCHAFFNER, Prof. K., Max-Planck-Institut für Strahlenchemie, Stiftstrasse 34-36, D-4330 Mülheim/Ruhr 1, Federal Republic of Germany
(TEL: +49 (208) 31073. TELEX: 856741 MPSTR D) (21)

SCHALLER, Dr. K. H., Institut für Arbeits- und Sozialmedizin und Poliklinik für Berufskrankheiten der Universität Erlangen-Nürnberg, Schillerstrasse 25-29, D-8520 Erlangen, Federal Republic of Germany (37)

SCHELZ, Dr. J. P., Research Center, Room E-1161, Johnson and Johnson Products Inc., US Route 1, New Brunswick, New Jersey 08903, USA (26)

SCHERAGA, Prof. H. A., Baker Laboratory, Department of Chemistry, Cornell University, Ithaca, New York 14853, USA
(TEL: +1 (607) 256 4034) (23)

SCHIEBER, Prof. M., School of Applied Science and Technology, Hebrew University of Jerusalem, Givat Ram, Jerusalem 91904, Israel
(TELEX: 25391 HU IL) (6)

SCHINDLER, Prof. P. W., Institut für Anorganische, Analytische und Physikalische Chemie, Universität Bern, Freiestrasse 3, CH-3000 Bern 9, Switzerland
(TEL: +41 (31) 65 42 41. TELEX: 33228 UNIBE CH) (30)

SCHMID, Dr. E. V., Höhenweg 13, CH-9000 St. Gallen, Switzerland (25)

SCHNEIDER, Prof. G. M., Lehrstuhl für Physikalische Chemie II, Ruhr-Universität Bochum, D-4630 Bochum 1, Federal Republic of Germany
(TEL: +49 (234) 7004249. TELEX: 0825860) (15)

SCHNEIDER, Dr. W. G., National Research Council of Canada, 100 Sussex Drive, Ottawa, Ontario K1A OR6, Canada
(TEL: +1 (613) 993 9821. TELEX: 533715 NRC OTT CA) (· , 10)

SCHOENFELD, Mr. R., Australian Journal of Chemistry, Commonwealth Scientific and Industrial Research Organization, PO Box 89, 314 Albert Street, East Melbourne, Victoria 3002, Australia
(TEL: +61 (3) 419 1333. TELEX: AA30236) (21)

SCHOLTE, Dr. Th. G., Central Laboratory, Dutch State Mines, PO Box 18, 6160 MD Geleen, Netherlands
(TEL: +31 (4494) 65605. TELEX: 56727 NL) (25)

SCHORSCH, Dr. G., Centre de Recherches d'Aubervilliers, Rhône Poulenc Industries, RP/RD/CRA, 12 rue des Gardinoux, F-93308 Aubervilliers, France
(TEL: +33 (1) 834 93 96. TELEX: 220306 RHÔNE F) (24)

SCHOUKENS, Dr. G., Laboratoire Central, Solvay & Cie SA, 310 Rue de Ransbeek, B-1120 Bruxelles, Belgium
(TEL: +32 (2) 267 38 30. TELEX: 26619 SOLVAY NOH BT B) (24)

SCHREIBER, Dr. B., Analytical Research and Development, Pharma Sandoz AG, Bau 360, CH-4002 Basel, Switzerland
(TEL: +41 (61) 24 42 63. TELEX: 63275 CH) (28)

SCHUTTE, Prof. C. J. H., Department of Chemistry, University of South Africa, PO Box 392, Pretoria 0001, Republic of South Africa
(TEL: +27 (12) 4402355. TELEX: 3777 SA) (17)

SCHUYFF, Prof. A., Laboratorium voor Algemene Chemie, Transitorium III, Universiteitscentrum "Uithof", Padualaan 8, Utrecht, Netherlands
(TEL: +31 (30) 533125) (15)

SCHWARTZ, Dr. M. K., Department of Biochemistry, Memorial Sloan-Kettering Cancer Center, 1275 York Avenue, New York, New York 10021, USA
(TEL: +1 (212) 794 5971) (36)

SCHWETLICK, Prof. K., Sektion Chemie der Technischen Universität Dresden, Mommsenstrasse 13, DDR-8027 Dresden, German Democratic Republic
(TEL: +37 (51) 4630. TELEX: 02278 DD) (21, 22)

SCROSATI, Prof. B., Dipartimento di Chimica, Università degli Studi di Roma, Piazzale Aldo Moro 5, I-00185 Roma, Italy
(TEL: +39 (6) 491769. TELEX: 613255 INFNRO I) (30)

SEFERIS, Prof. J. C., Polymeric Composites Laboratory, Department of Chemical Engineering, BF-10, University of Washington, Seattle, Washington 98195, USA
(TEL: +1 (206) 543 2250/9371) (24)

SEIBER, Prof. J., Department of Environmental Toxicology, University of California, Davis, California 95616, USA (33)

SENDA, Prof. M., Department of Agricultural Chemistry, Faculty of Agriculture, Kyoto University, Yoshida-machi, Sakyo-ku, Kyoto, 606 Japan
(TEL: +81 (75) 751 2111. TELEX: 5422693 LIBKYU J) (29)

SÉNÉMAUD, Dr. CHRISTIANE, Laboratoire de Chimie Physique, Université de Paris VI, 11 rue Pierre et Marie Curie, F-75231 Paris Cédex 05, France
(TEL: +33 (1) 336 28 89/39 01) (28)

SENVAR, Prof. C. B., Mühendislik Fakültesi, Fen Fakültesi, Istanbul, Turkey (12)

SERSALE, Prof. R., Istituto di Chimica Applicata, Università di Napoli, Piazzale Tecchio, I-80125 Napoli, Italy
(TEL: +39 (81) 624238/631600. TELEX: 7691271 I) (20)

SETTON, Dr. R., Laboratoire d'Etudes Physico-Chimiques, Centre National de la Recherches Scientifique, F-45045 Orléans, France
(TEL: +33 (38) 63 22 16) (20)

SHANI, Prof. A., Department of Chemistry, Ben-Gurion University of the Negev, Beer-Sheva 84120, Israel
(TEL: +972 (57) 664398/391. TELEX: 5253 UNASI IL) (12)

SHAPCOTT, Dr. D., Département de Pédiatrie, Faculté de Médecine, Université de Sherbrooke, Sherbrooke, Québec J1H 5N4, Canada
(TELEX: 05-836184 CA) (36)

SHARP, Dr. D. B., Environmental Science/Formulations, Monsanto Agricultural Products Co., 800 North Lindbergh Boulevard, St. Louis, Missouri 63166, USA
(TELEX: 44-7282) (33)

SHAW, Prof. D. G., Institute of Marine Science, University of Alaska, Fairbanks, Alaska 99701, USA
(TEL: +1 (907) 474 7723) (30)

SHEASLEY, Dr. D., Analytical Research Laboratories, Rohm and Haas Co., POB 219, Bristol, Pennsylvania 19007, USA (26)

SHEPPARD, Prof. N., School of Chemical Sciences, University of East Anglia, University Plain, Norwich NR4 7TJ, UK
(TEL: +44 (603) 56161. TELEX: 97154 G) (15)

SHI, Prof. L., Institute of Chemistry, Chinese Academy of Sciences, Chung Kuan Tsun, Beijing, China
(TEL: 282281) (24)

SHIBATA, Dr. S., Government Industrial Research Institute, Nagoya, Agency of Industrial Science and Technology, Ministry of International Trade and Industry, Hirate-machi, Kita-ku, Nagoya, 462 Japan
(TEL: +81 (52) 911 2111) (28)

SHIMA, Dr. MASAKO, National Science Museum, 17-20 Ueno Park, Taito-ku, Tokyo, 110 Japan
(TEL: +81 (3) 822 0111) (19)

SHIMIZU, Prof. J., Tokyo Institute of Technology, Ookayama, Meguro-ku, Tokyo, 152 Japan
(TEL: +81 (3) 72 6111. TELEX: 2466360 TITECH J) (24)

SHIMOZAWA, Prof. J. T., Faculty of Science and Engineering, Saitama University, 255 Shimo Okubo, Urawa, 338 Japan
(TEL: +81 (488) 52 2111) (11)

SHONHORN, Dr. H., Bell Laboratories, 600 Mountain Avenue, Murray Hill, New Jersey 07974, USA
(TEL: +1 (201) 582 3000. TELEX: 13-8650) (26)

SHORLAND, Dr. F. B., Department of Biochemistry, Victoria University of Wellington, Private Bag, Wellington, New Zealand
(TEL: +64 (4) 721000) (33)

SHOUKRY, Dr. A. EL-SAWI, National Organization for Drug Control and Research, POB 29, Cairo, Arab Republic of Egypt
(TEL: 850005) (10)

SIANESI, Prof. D., Montedison, Coordinamento Strategico, Foro Bonaparte 31, I-20100 Milano, Italy (10)

SIBILIA, Dr. J. P., Analytical Sciences Laboratory, Allied Chemical Corpn., Corporate Headquarters, POB 1021R, Morristown, New Jersey 07960, USA
(TELEX: 136410, 136470) (26)

SICKFELD, Dr. J., Bundesanstalt für Materialprüfung (BAM), Unter den Eichen 87, D-1000 Berlin 45 (Dahlem), Federal Republic of Germany
(TEL: +49 (30) 81043140. TELEX: 183261 BAMB D) (25)

SIDHU, Dr. G. S., Council of Scientific and Industrial Research, Rafi Marg, New Delhi-110001, India
(TEL: +91 (11) 383652. TELEX: ND-2522 ABC CSIR IN) (10)

SIEKIERSKI, Prof. S., Department of Nuclear Chemistry, Institute of Nuclear Research, Dorodna 16, PL-03 195 Warsaw, Poland (31)

SIEMION, Prof. I. Z., Instytut Chemii, Uniwersytet Wrocławski, ul. Joliot-Curie 14, PL-50 383 Wrocław, Poland
(TELEX: 0712791 PL) (12)

SIGGAARD-ANDERSEN, Prof. O., Department of Clinical Chemistry, Københavns Amtssygehus i Herlev, Herlev Ringvej, DK-2730 Herlev, Denmark
(TEL: +45 (2) 94 27 33) (36)

SIGWALT, Prof. P., Laboratoire de Chimie Macromoléculaire, Université Pierre et Marie Curie, Tour 44, 4 place Jussieu, F-75230 Paris Cédex 05, France
(TEL: +33 (1) 329 12 21) (22, 23, 24)

SIKYTA, Dr. B., Mikrobiologický Ústav, Ceskoslovenská Akademie Ved, Vídenská 270/1083, CS-142 20 Praha 4-Krc, Czechoslovakia
(TEL: +42 (2) 4713221. TELEX: 121341 C) (32)

SILBERBERG, Prof. A., Department of Polymer Research, Weizmann Institute of Science, POB 26, Rehovot 76100, Israel
(TEL: +972 (54) 82111/83323. TELEX: 361900 WIX IL) (23)

SIMAK, Dr., BASF Aktiengesellschaft, D-6700 Ludwigshafen/Rhein, Federal Republic of Germany
(TEL: +49 (621) 606479. TELEX: 464811 BASF ZENTRALE D) (25)

SIMEON, Dr. V., Institut za Medicinska Istraživanja i Medicinu Rada, POB 291, ul. Moše Pijade 158, YU-41001 Zagreb, Yugoslavia
(TEL: +38 (41) 434188) (19)

SIMIONESCU, Prof. C. I., Akademia Republicii Socialiste România, Strada Universitatii Nr. 16, Iasi, Romania
(TELEX: 22370 R) (12, 23)

SIMON, Prof. W., Laboratorium für Organische Chemie der Eidgenössischen Technischen Hochschule Zürich, Universitätstrasse 16, CH-8092 Zürich, Switzerland
(TEL: +41 (1) 256 28 98) (28)

SING, Prof. K. S. W., Department of Chemistry, Brunel University, Uxbridge, Middlesex UB8 3PH, UK
(TEL: +44 (895) 37188. TELEX: 261173 G) (11, 17, 18)

VAN SITTERT, Mr. N. J., Shell Nederland-MDH 33, Postbus 162, 2501 Den Haag, Netherlands
(TEL: +31 (70) 774805. TELEX: 31005 NL) (37)

SJÖHOLM, Dr. J. A. A., Höganäs Koncernen AB, Foretagshalsovarnen, Box 501, S-263 01 Höganäs, Sweden
(TEL: +46 (42) 4238226) (37)

SKAGIUS, Doc. K., AB Fortia, Box 604, S-751 25 Uppsala 1, Sweden
(TEL: +46 (18) 163000. TELEX: 76027 PHARMUP S) (10)

SKRZECZ, Dr. A., Instytut Chemii Fizycznej, Polskiej Akademii Nauk, ul. Kasprzaka 44-52, PL-01 224 Warszawa, Poland
(TEL: +48 (22) 328392. TELEX: 817097 ICHF PL) (30)

SLOAN, Mr. T. E., Chemical Abstracts Service, POB 3012, 2540 Olentangy River Road, Columbus, Ohio 43210, USA
(TEL: +1 (614) 421 3600. TELEX: 810 482 1608 CHEM ABS COL) (19)

SMITH, Dr. A. L., Spectroscopy Laboratory, Dow Corning Corpn., POB 1592, Midland, Michigan 48640, USA (17)

SMITH, Mr. N. J., Occupational Medicine and Hygiene Laboratories, 403 Edgware Road, London NW2 6LN, UK
(TEL: +44 (1) 450 8911) (37)

SMITH, Prof. P. A. S., Department of Chemistry, University of Michigan, Ann Arbor, Michigan 48109, USA
(TEL: +1 (313) 764 7355. TELEX: 8102236056 U OF M AA) (21)

SMYTH, Dr. W. F., Norbrook Laboratories Ltd., Station Works, Newry BT35 6JP, Northern Ireland, UK
(TEL: +0044 (0693) 4435. TELEX: 747048 G) (29)

SOIGNET, Dr. D. M., Natural Polymers Research Laboratory, US Department of Agriculture, Federal Research Southern Region, Southern Regional Research Center, POB 19687, 1100 Robert E. Lee Boulevard, New Orleans, Louisiana 70179, USA (26)

SOKOLOVSKAYA, Prof. EVDOKIYA M., Faculty of Chemistry, Lomonosov Moscow State University, Leninskie Gory, SU-117234 Moscow, V-234, USSR
(TEL: +7 (095) 139 1596) (11)

SOLOMON, Dr. D. H., Division of Applied Organic Chemistry, Commonwealth Scientific and Industrial Research Organization, PO Box 4331, Melbourne, Victoria 3001, Australia (25)

SOMMER, Prof. L., Katedra Analytické Chemie, Universita J. E. Purkyně, Kotlářská 2, CS-611 37 Brno, Czechoslovakia
(TEL: +42 (5) 51112) (27)

SOMSEN, Prof. G., Scheikundig Laboratorium der Vrije Universiteit, De Boelelaan 1083, 1081 HV Amsterdam, Netherlands
(TEL: +31 (20) 5485364. TELEX: 11329 NL) (16)

SÖRBO, Prof. B., Klinisk-Kemiska Laboratoriet, Regionsjukhuset, Universitet i Linköping, S-581 85 Linköping, Sweden
(TEL: +46 (13) 193220) (37)

SÖRVIK, Dr. E., Polymergruppen, Chalmers Tekniska Högskola, Göteborg Universitet, S-412 96 Göteborg, Sweden
(TEL: +46 (31) 810100. TELEX: 2369 CHERLBIB S) (25)

SOTIROPOULOU, Dr. G., Z. Pigis 43, Kato Halandri, Athens 152 31, Greece (36)

SPERLING, Dr. K. R., Biologische Anstalt Helgoland, Labor Sülldorf, Wüstland 2, D-2000 Hamburg 55, Federal Republic of Germany
(TEL: +49 (40) 871026) (37)

SPOOR, Dr. H., BASF Aktiengesellschaft, D-6700 Ludwigshafen/Rhein, Federal Republic of Germany
(TELEX: 464811 BASF D) (25)

STAMM, Prof. D., Max-Planck-Institut für Psychiatrie, Abteilung für Klinische Chemie, Kraepelinstrasse 10, D-8000 München 40, Federal Republic of Germany
(TEL: +49 (89) 39102270) (36)

STARCK, Mr. P., Neste Oy, Polymer Plant, SF-06850 Kulloo, Finland
(TEL: +358 (15) 692791. TELEX: 1721 SF) (24, 25)

STARITA, Dr. J. M., Rheometrics Inc., 1 Possumtown Road, Piscataway, New Jersey 08854, USA
(TEL: +1 (201) 560 8550. TELEX: 219107) (24)

START, Mr. P. A., Department of Chemistry, University College, Dublin, Ireland
(TEL: +0044 (0001) 693244) (12)

STARÝ, Prof. J., Katedra Jaderné Chemie, Fakulta Jaderná a Fyzikálně Inženýrská, České Vysoká Ucení Technické v Praze, Břehova ul. 7, CS-115 19 Praha 1-Staré Město, Czechoslovakia
(TEL: +42 (2) 542320) (28)

STENIUS, Prof. P., Swedish Institute for Surface Chemistry, Box 5607, S-114 86 Stockholm, Sweden
(TEL: +46 (8) 222545. TELEX: 14374 STURES) (17)

STEPHAN, Prof. K., Universität Stuttgart, Pfaffenwaldring 9, D-7000 Stuttgart 80, Federal Republic of Germany
(TEL: +49 (711) 6856102/3. TELEX: 07-255445 D) (16)

STEPHEN, Dr. W. I., Haworth Building, Department of Chemistry, University of Birmingham, POB 363, Birmingham B15 2TT, UK
(TEL: +44 (21) 472 1301) (27)

STERN, Dr. R. M., Working Environment Research Group, Danish Welding Institute, Park Allee 345, DK-2600 Glostrup, Denmark
(TEL: +45 (2) 96 88 00) (37)

STERNBERG, Dr. Z. W., Institute Ruder Boskovic, POB 1016, YU-41001 Zagreb, Yugoslavia (15)

STEUER, Prof. W., Medizinisches Landesuntersuchungsamt, Postfach 2642, D-7000 Stuttgart 1, Federal Republic of Germany
(TEL: +49 (711) 2023-1) (38)

STEVENS, Dr. B., Department of Applied Biology, Royal Melbourne Institute of Technology, Latrobe Street, Melbourne, Victoria 3001, Australia (38)

STEWART, Dr. G. G., Production-Research Department, Labatt Brewing Co. Ltd., 150 Simcoe Street, London, Ontario N6A 4M3, Canada
(TEL: +1 (519) 673 5326. TELEX: 064 7244 CA) (32)

STEYN, Dr. P. S., National Chemical Research Laboratory, Council for Scientific and Industrial Research, PO Box 395, Pretoria 0001, Republic of South Africa
(TEL: +27 (12) 869211. TELEX: 3-630 SA) (32)

STICKLER, Dr. M., Röhm GmbH, Chemische Fabrik, Postfach 4242, D-6100 Darmstadt, Federal Republic of Germany
(TEL: +49 (6151) 184537. TELEX: 419485 R D) (26)

STOEPPLER, Dr. M., Institut für Chemie der Kernforschungsanlage Jülich GmbH, Institut 4: Angewandte Physikalische Chemie, Postfach 1913, D-5170 Jülich 1, Federal Republic of Germany
(TEL: +49 (2461) 613179. TELEX: 833 556 KFA D) (36, 37, 38)

STOICA, Dr. RODICA, Surfactants Department, Institutul de Cercetări Chimice (ICECHIM), 202 Splaiul Independenţei, sect. 6, R-77208 Bucureşti, Romania
(TEL: 159535. TELEX: 10944 ICECHIM R) (33)

ST. PIERRE, Prof. L. E., Department of Chemistry, Otto Maass Chemistry Building, McGill University, 801 Sherbrooke Street West, Montreal, Québec H3A 2K6, Canada
(TEL: +1 (514) 392 4469) (23)

STRAND, Mrs. SISSEL, Dr. V. Fürst Medisinsk Laboratorium, Søren Bulls Vei 25, Oslo 10, Norway
(TEL: +47 (2) 303730) (35)

STRASHEIM, Dr. A., Chemistry Department, University of Pretoria, Pretoria 0002, Republic of South Africa
(TEL: +27 (12) 436051) (28)

ŠTRAUS, Prof. B., Zavod za Medicinsku Biokemiju, Farmaceutsko-Biokemijski Fakultet, Sveučilište u Zagrebu, Domajojeva 2, YU-41000 Zagreb, Yugoslavia
(TEL: +38 (41) 4146) (36)

STRAUSZ, Prof. O. P., Department of Chemistry, University of Alberta, Edmonton, Alberta T6G 2G2, Canada
(TEL: +1 (403) 432 3801. TELEX: 037 2979 CA) (22)

STRAZIELLE, Dr. C., Centre de Recherches sur les Macromolécules du Centre National de la Recherche Scientifique, 6 rue Boussingault, F-67083 Strasbourg Cédex, France (25)

STREHLOW, Dr. C. D., Westminster Children's Hospital, Vincent Square, London SW1P 2NS, UK (38)

STROM, Prof. R., Cattedra di Chimica e Microscopia Clinica, Università "La Sapienza", Piazzale Aldo Moro 5, I-00185 Roma RM, Italy (36)

STUMPP, Prof. E., Anorganisch-Chemisches Institut, Technische Universität Clausthal - Lehrstuhl A, D-3392 Clausthal-Zellerfeld, Federal Republic of Germany
(TEL: +49 (5323) 721. TELEX: 953828 TU CLZ D) (20)

SUBRAMANIAM, Dr. A., Rubber Research Institute of Malaysia, c/o POB 150, Kuala Lumpur, Malaysia
(TEL: +60 (3) 467033. TELEX: 30369 RRIM MA) (24)

SUBRAMANIAN, Dr. K. S., Environmental Health Centre, Health and Welfare, Tunney's Pasture, Ottawa, Ontario K1A 0L2, Canada
(TEL: +1 (613) 992 1388) (38)

SUGA, Prof. H., Chemical Thermodynamics Laboratory, Department of Chemistry, Faculty of Science, University of Osaka, 1-1 Machikaneyama, Toyonaka, Osaka, 560 Japan
(TEL: +81 (6) 844 1151) (13, 16)

SUKH DEV, Dr., Malti-Chem Research Center, Nandesari, Baroda-391340, India
(TEL: Nandesari 225. TELEX: 0175-396 ANU IN) (21)

SULTAN, Mr. S. Å., Polymergruppen, Chalmers Tekniska Högskola, Göteborg Universitet, S-412 96 Göteborg, Sweden
(TEL: +46 (31) 810100. TELEX: 2369 CHERLBIB S) (25)

SUNDELÖF, Prof. L.-O., Institute of Inorganic and Physical Chemistry, University of Uppsala, Biomedicum, Box 574, S-751 23 Uppsala, Sweden (24)

SUNDERMAN, Jr., Prof. F. W., Departments of Laboratory Medicine and Pharmacology, University of Connecticut Medical School, 263 Farmington Avenue, Farmington, Connecticut 06032, USA
(TEL: +1 (203) 674 2328) (10, 13, 14, 35, 37)

SUNDERMAN, Sr., Prof. F. W., Institute for Clinical Science, Hahnemann Medical College, 230 North Broad Street, Philadelphia, Pennsylvania 19102, USA
(TEL: +1 (215) 568 6866) (37)

SUTER, Dr. U. W., Department of Chemical Engineering, Room 66-456, Massachusetts Institute of Technology, Cambridge, Massachusetts 02139, USA
(TEL: +1 (617) 253 4590. TELEX: 92 1473 MITCAM) (23)

SVEHLA, Dr. G., Department of Chemistry, David Keir Building, Queen's University of Belfast, Stranmillis Road, Belfast BT9 5AG, Northern Ireland, UK
(TEL: +0044 (0232) 661111. TELEX: 74487 G) (28)

SZAFRANSKI, Dr. A., Department NB-6/3, Institute of Industrial Chemistry, ul. Rydygiera 8, PL-01 793 Warsaw, Poland
(TEL: +48 (22) 388021. TELEX: 813586 ICH PL) (30)

SZÁNTAY, Prof. C., Institute of Organic Chemistry, Technical University of Budapest, Gellért tér 4, H-1521 Budapest, Hungary
(TEL: +36 (1) 450 300. TELEX: 225931 MÜEGY H) (11)

TAKEDA, Dr. K., Shionogi Research Laboratories, Shionogi & Co. Ltd., Sagisu 5-12-4, Fukushima-ku, Osaka, 553 Japan
(TEL: +81 (6) 548 5861. TELEX: 63596 SHIOGI J) (10)

TALIERI-GIANOPOULOU, Dr. MAROULIO J., Nea Penteli, Athens 152 36, Greece
(TEL: +30 (1) 8044551) (35)

TAL'ROSE, Prof. V. I., Institute of Chemical Physics, Academy of Sciences of USSR, Kosygin Street 4, SU-117977, SZD-1, Moscow, V-334, USSR (17)

TAMARU, Prof. K., Department of Chemistry, Faculty of Science, University of Tokyo, 3-1 Hongo 7-chome, Bunkyo-ku, Tokyo, 113 Japan
(TEL: +81 (3) 812 2111. TELEX: 25510 UNITOKYO J) (15)

TAN, Dr. A. S., Rubber Research Institute of Malaysia, c/o POB 150, Kuala Lumpur, Malaysia
(TEL: +60 (3) 467033. TELEX: 30369 RRIM MA) (24)

TAN, Dr. Y. Y., Laboratorium voor Polymeerchemie RUG, Nyenborgh 16, 9747 AG Gröningen, Netherlands
(TEL: +31 (50) 117203. TELEX: 53410 RUGRO NL) (26)

TANAKA, Dr. N., Nippon Jimuki Co. Ltd., 12, 3-chome, Kanda Ogawa-cho, Chiyoda-ku, Tokyo, 101 Japan
(TEL: +81 (3) 292 1511) (29)

TANDON, Dr. S. K., Industrial Toxicology Research Centre, POB 80, Mahatma Gandhi Marg, Lucknow-226001, India
(TEL: +91 (227) 28228) (37)

TASUMI, Prof. M., Department of Chemistry, Faculty of Science, University of Tokyo, 3-1 Hongo 7-chome, Bunkyo-ku, Tokyo, 113 Japan
(TEL: +81 (3) 812 2111. TELEX: 33659 UTYOSCI J) (17)

TAWN, Mr. A. R. H., Coates Brothers and Co. Ltd., St. Mary Cray, Orpington, Kent BR5 3PP, UK
(TEL: +44 (689) 27080) (25)

TAYLOR, Dr. A., Department of Biochemistry, University of Surrey, Guildford, Surrey GU2 5XH, UK (38)

TEMPLE, Dr. W. A., National Poisons/Hazardous Chemical Information Centre, Otago Medical School, POB 913, Dunedin, New Zealand (36)

TERENT'EVA, Dr. EUGENIA A., Institute of Organoelement Compounds, Academy of Sciences of USSR, ul. Vavilov 28, SU-117813 Moscow, B-334, USSR
(TEL: +7 (095) 135 9218) (28)

TER-MINASSIAN-SARAGA, Dr. LISBETH, Laboratoire de Physico-Chimie des Surfaces et des Membranes du Centre National de la Recherche Scientifique, UER, Biomédicale 81, 45 rue des Saints-Pères, F-75270 Paris Cédex 06, France
(TEL: +33 (1) 297 53 18) (17)

TEUPEL, Dr. M., Henkel KGaA, Postfach 1100, D-4000 Düsseldorf 1, Federal Republic of Germany
(TEL: +49 (211) 7974349. TELEX: 085817-0 DÜSSELDORF D) (33)

THOMA, Prof. H., Instituto de Química, Universidade São Paulo, Caixa Postal 20780, 01000 São Paulo - SP, Brazil (27)

THOMASSEN, Dr. Y., Institute of Occupational Health, Gydas Vei 8, POB 8149 Dep., Oslo 1, Norway
(TEL: +47 (2) 466850) (36, 37, 38)

THOMPSON, Dr. H., Kent Industrial Measurements Ltd., EIL Analytical Instruments, Hanworth Lane, Chertsey, Surrey KT16 9LF, UK
(TEL: +44 (9328) 62671. TELEX: 264022 EILCHT G) (29)

THOMSON, Dr. CHRISTINE D., Department of Nutrition, University of Otago, POB 56, Dunedin, New Zealand
(TEL: +64 (24) 771640) (38)

THOR, Mr. A., Secretariat of ISO/TC 12, SIS-Standardiseringskommissionen i Sverige, Box 3 295, S-103 66 Stockholm, Sweden (12)

THORBURN BURNS, Prof. D., Department of Chemistry (Analytical), David Keir Building, Queen's University of Belfast, Stranmillis Road, Belfast BT9 5AG, Northern Ireland, UK
(TEL: +0044 (0232) 661111. TELEX: 74487 G) (27)

TISCORNIA, Prof. E., Istituto Chimica Formaceutica Università, Viale Benedetto XV/3, I-16132 Genova, Italy (33)

TITOV, Dr. V. N., Laboratory of Clinical Chemistry, All Union Cardiological Research Centre, Academy of Medical Sciences USSR, 3rd Cherepcovsky Street 15-a, Kunzevo, SU-122500 Moscow, USSR (35, 36)

TJALVE, Dr. H., Department of Toxicology, University of Uppsala, Box 573, S-751 23 Uppsala, Sweden
(TEL: +46 (18) 152000) (37)

TODD, Dr. J. F. J., University Chemical Laboratory, University of Kent at Canterbury, Canterbury, Kent CT2 7NH, UK
(TEL: +44 (227) 66822. TELEX: 965449 G) (17)

TOKUMARU, Prof. K., Department of Chemistry, University of Tsukuba, Sakura-mura, Niihari-gun, Ibaraki, 305 Japan
(TEL: +81 (298) 53 4315. TELEX: 365 2580 UNTUKU J) (22)

TÖLG, Prof. G., Institut für Spektrochemie und Angewandte Spektroskopie, Postfach 778, Bunsen-Kirchhoff Strasse 11, D-4600 Dortmund 1, Federal Republic of Germany
(TEL: +49 (231) 129001/04) (27)

TOMKINS, Prof. R. P. T., Department of Chemical Engineering and Chemistry, New Jersey Institute of Technology, 323 High Street, Newark, New Jersey 07102, USA
(TEL: +1 (201) 645 5060) (30)

TOMLINSON, Prof. J. W., Department of Chemistry, Victoria University of Wellington, Private Bag, Wellington, New Zealand
(TEL: +64 (4) 721000) (16)

TONKS, Prof. D. B., Division of Clinical Chemistry, Montreal General Hospital, 1650 Cedar Avenue, Montreal, Québec H3G 1A4, Canada
(TEL: +1 (514) 937 6011) (10, 11, 36, 37, 38)

TOTH, Dr. A., Société Astral, BP 4, 16 rue de Déportés, F-60160 Montataire, France (25)

TÓTH, Dr. I., Magyar Tudományos Akadémia, Központi Kémiai Kutato Intézete, Pusztaszeri utca 59-67, H-1525 Budapest, Hungary
(TELEX: 226686 KEMIA H) (22)

TÓTH, Dr. KLÁRA, Altalános és Analitikai Kémiai Tanszék, Budapesti Müszaki Egyetem, Gellért tér 4, H-1521 Budapest, Hungary
(TEL: +36 (1) 453 500. TELEX: 225931 MÜEGY H) (29)

TOULLEC, Dr. J., Laboratoire de Chimie Organique Physique, Institut de Topologie et de Dynamique des Systèmes, Centre National de la Recherche Scientifique, Université de Paris VII, 1 rue Guy de la Brosse, F-75005 Paris, France
(TEL: +33 (1) 336 25 25) (21, 22)

TOUSSAINT, Mr. A., Laboratoire IVP, Avenue Pierre Holoffe, B-1342 Limelette, Belgium (25)

TOWNSHEND, Dr. A., Department of Chemistry, University of Hull, Hull HU6 7RX, UK
(TEL: +44 (482) 497027. TELEX: 52530 UNIHUL G) (27)

TRAPPENIERS, Prof. N. J., Van der Waals Laboratorium, Universiteit van Amsterdam, Valckenierstraat 67, 1018 XE Amsterdam, Netherlands (5)

TRASATTI, Prof. S., Dipartimento di Chimica Fisica ed Elettrochimica, Via Venezian 21, I-20133 Milano, Italy
(TEL: +39 (2) 293516. TELEX: 320484 UNI MI I) (16)

TREINDL, Prof. L., Katedra Fysikální Chemie, Přírodovědecka Fakulta UK, Mlynská dolina, CS-842 15 Bratislava, Czechoslovakia
(TEL: +42 (7) 325501. TELEX: -) (15)

TREINER, Dr. C., Bâtiment F, Laboratoire d'Electrochimie, Université Paris VI, 4 place Jussieu, F-75005 Paris, France
(TEL: +33 (1) 336 25 25) (30)

TSCHOEGL, Prof. N. W., Division of Chemistry and Chemical Engineering, California Institute of Technology, Pasadena, California 91125, USA (6)

TSUCHIYA, Prof. M., Department of Industrial Chemistry, Faculty of Engineering, University of Tokyo, 3-1 Hongo 7-chome, Bunkyo-ku, Tokyo, 113 Japan
(TEL: +81 (3) 812 2111. TELEX: 25510 UNITOKYO J) (17)

TSUZUKU, Prof. T., Physical Science Laboratories, Nihom University, 7-24-1 Narashinodai, Funabashi, 274 Japan (20)

TUCK, Prof. D. G., Department of Chemistry, University of Windsor, Windsor, Ontario N9B 3P4, Canada
(TEL: +1 (519) 253 4232. TELEX: 06477684 CA) (29, 30)

TULLEY, Dr. R. T., Department of Health and Human Resources, Earl K. Long Memorial Hospital, POB 52999, 5825 Airline Highway, Baton Rouge, Louisiana 70805, USA (38)

TULLOCH, Dr. A. P., Plant Biotechnology Institute, National Research Council of Canada, University Campus, 110 Gymnasium Road, Saskatoon, Saskatchewan S7N OW9, Canada
(TEL: +1 (306) 665 5258) (33)

TURI, Dr. EDITH A., Research Center, Allied Chemical Corpn., Corporate Headquarters, POB 1021R, Morristown, New Jersey 07960, USA
(TEL: +1 (201) 455 2862. TELEX: 136410, 136470) (26)

TURNER, Dr. D. W., Physical Chemistry Laboratory, University of Oxford, South Parks Road, Oxford OX1 3QZ, UK
(TEL: +44 (865) 53322. TELEX: 8368 BIOXFO G) (13)

UBBELOHDE, Prof. A. R., Department of Chemical Engineering and Chemical Technology, Imperial College of Science and Technology, Prince Consort Road, South Kensington, London SW7 2BY, UK
(TEL: +44 (1) 5895111. TELEX: 261503 G) (20)

UCHIDA, Dr. M., Projects Development Department, Teijin Ltd., Iino Building, 1-1 Uchisaiwai-cho, 2-chome, Chiyoda-ku, Tokyo, 100 Japan
(TEL: +81 (3) 506 4111/4136. TELEX: 2223569 TEIJIN J) (10, 24)

UDEN, Prof. P. C., Department of Chemistry, GRC Tower A, University of Massachusetts, Amherst, Massachusetts 01003, USA
(TELEX: 95 5491) (27)

UENO, Prof. K., Department of Industrial Chemistry, Kumamoto Institute of Technology, Ikeda 4-22-1, Kumamoto, 860 Japan
(TEL: +81 (963) 3111. TELEX: 7627-76 DOJINL J) (27)

UNBEHEND, Dr. M., Unilever Forschungsgesellschaft mbH, Postfach 1568, Behringstrasse 154, D-2000 Hamburg 50, Federal Republic of Germany
(TELEX: 2164293 UNAT D) (25)

URE, Dr. A. M., Department of Spectrochemistry, Macaulay Institute for Soil Research, Craigiebuckler, Aberdeen AB9 2QJ, UK
(TEL: +44 (224) 38611. TELEX: 73458 UNIABN G) (28)

VAESSEN, Dr. H. A. M. G., Laboratory for Chemical Analysis of Foodstuffs, Rijksinstituut voor de Volksgezondheid, Postbus 1, 3720 BA Bilthoven, Netherlands
(TEL: +31 (30) 742339. TELEX: 47215 RIVBH NL) (32)

VALDIGUIÉ, Prof. P., Centre Hospitalier Régional de Toulouse, CHU Toulouse-Rangueil, Chemin du Vallon, F-31054 Toulouse Cédex, France
(TEL: +33 (61) 53 11 33. TELEX: 530217 CHRT DG F) (35)

VALENTA, Dr. P., Institut für Chemie der Kernforschungsanlage Jülich GmbH, Institut 4: Angewandte Physikalische Chemie, Postfach 1913, D-5170 Jülich 1, Federal Republic of Germany
(TEL: +49 (2461) 613177. TELEX: 833 556 KFA D) (29)

VAN BRANDT, Prof. P., Laboratoire de Chimie Générale et Organique, BP 71, Place Louis Pasteur 1, B-1348 Louvain-la-Neuve, Belgium (21, 22)

VANCSÓ-SZMERCSÁNYI, Dr. I., Müanyagipari Kutató Intézete, Hungária krt. 114, H-1950 Budapest, Hungary
(TEL: +36 (1) 634 200. TELEX: 226493 MUEKI H) (26)

VANDERDEELEN, Dr. J., Physical and Radiobiological Chemistry, Faculty of Agriculture, State University Gent, Coupure Links 653, B-9000 Gent, Belgium
(TEL: +32 (91) 23 69 61. TELEX: 12754 RUGENT B) (30)

VAN GRIEKEN, Prof. R. E., Departement Scheikunde, Universitaire Instelling Antwerpen, Universiteitsplein 1, B-2610 Wilrijk, Belgium
(TEL: +32 (31) 828 25 28. TELEX: 33646 B) (28)

VANNERBERG, Prof. N. G., Department of Inorganic Chemistry, Chalmers University of Technology, S-412 96 Göteborg, Sweden
(TEL: +46 (31) 810100. TELEX: 2369 CHERLBIB S) (12)

VASSILATOS, Dr. G., Engineering Technology Laboratory, E. I. du Pont de Nemours & Co. Inc., Experimental Station, Wilmington, Delaware 19898, USA
(TEL: +1 (302) 772 3280. TELEX: 835420 DUPONT) (24)

VEILLON, Dr. C., Vitamin and Mineral Nutrition Laboratory, Human Nutrition Research Center, Building 307, Room 215, Beltsville, Maryland 20705, USA
(TEL: +1 (301) 344 2010) (38)

VELICK, Prof. S. F., 201 South Biology Building, University of Utah, Salt Lake City, Utah 84112, USA
(TELEX: 910 925 5283 UNIV UTAH SLC) (13)

VENKATARAMAN, Prof. B., Chemical Physics Division, Tata Institute of Fundamental Research, Homi Bhabha Road, Bombay-400005, India (11)

VEPREK, Dr. S., Anorganisch-Chemisches Institut der Universität Zürich, Winterthurerstrasse 190, CH-8057 Zürich, Switzerland
(TEL: +41 (1) 257 11 11. TELEX: 55575 UNIZI CH) (15)

DE VERDIER, Prof. C.-H., Klinisk-Kemiska Centrallaboratoriet, Akademiska Sjukhuset, S-751 85 Uppsala, Sweden
(TEL: +46 (18) 166372) (36)

VERHOEVEN, Prof. J. W., Laboratorium voor Organische Scheikunde der Universiteit van Amsterdam, Nieuwe Achtergracht 129, 1018 WS Amsterdam, Netherlands
(TEL: +31 (20) 5222676. TELEX: 16460 NL) (22)

VERLINDEN, Dr. MARLEEN, c/o Janssen Pharmaceutica, Department of Clinical Research, Turnhoutseweg 30, B-2340 Beerse, Belgium
(TEL: +32 (14) 60 21 66. TELEX: 32540 B) (38)

VERNERKAR, Prof. R., Department of Inorganic and Physical Chemistry, Indian Institute of Science, Malleswaram, Bangalore-560012, India
(TELEX: 0845-8349 IISC IN) (20)

VERSIECK, Prof. J., Department of Internal Medicine, Division of Gastroenterology, University Hospital, De Pintelaan 185, B-9000 Gent, Belgium (38)

VESALA, Dr. A., Department of Chemistry and Biochemistry, University of Turku, Vesilinnantie 5, SF-20500 Turku 50, Finland
(TEL: +358 (21) 645736) (30)

VESTERBERG, Prof. O., Occupational Health Department, Chemistry Division, National Board of Occupational Safety and Health, S-171 84 Solna, Sweden
(TEL: +46 (8) 7309000) (38)

VIDAL, Dr. J. L., L PC Naphtachimie, BP 12, F-13117 Lavera, France (25)

VIHKO, Prof. R., Department of Clinical Chemistry, University of Oulu, SF-90220 Oulu 22, Finland
(TEL: +358 (81) 332033. TELEX: 32305 OYKS SF) (35, 36)

VINCENT, Dr. J. A. J. M., Central Laboratory, Dutch State Mines, PO Box 18, 6160 MD Geleen, Netherlands
(TEL: +31 (4494) 65855. TELEX: 56727 NL) (25)

VISAPÄÄ, Mrs. ANNELI, Finnish Customs Laboratory, Tullilaboratorio, Tekniikantie 13, SF-02150 Espoo 15, Finland
(TEL: +358 (0) 4550311. TELEX: 121559 TULHS SF) (33)

VISSER, Prof. B. F., Ziekenhuis St. Annadal, Postbus 1918, 6201 BX Maastricht, Netherlands
(TEL: +31 (43) 862691. TELEX: 56745 ANDAL NL) (36)

VLČEK, Prof. A. A., Ústav Fyzikální Chemie a Elektrochemie J Heyrovského, Československá Akademie Věd, Vlašská 9, CS-118 40 Praha 1-Malá Strana, Czechoslovakia
(TEL: +42 (2) 534150) (10, 29)

VLIEGENTHART, Prof. J. F. G., Department of Bio-Organic Chemistry, State University Utrecht, Croesestraat 79, 3522 AD Utrecht, Netherlands
(TEL: +31 (30) 882311) (13)

VONDERSCHMITT, Prof. D. J., Medizinisch-Chemisches Zentrallabor, Universitätsspital, CH-8091 Zürich, Switzerland
(TEL: +41 (1) 255 22 60) (35)

VONK, Dr. J. W., Organisch Chemisch Instituut TNO, Postbus 5009, 3502 JA Utrecht, Netherlands
(TEL: +31 (30) 882721) (33)

VORSTER, Dr. O., Division of Chemical Sciences, Technikon Pretoria, Kerkstraat 420, Pretoria 0002, Republic of South Africa (25)

DE VRIES, Dr. A. J., Centre de Recherches d'Aubervilliers, Rhône Poulenc Industries, 12 rue des Gardinoux, F-93308 Aubervilliers Cédex, France
(TEL: +33 (1) 834 93 96. TELEX: 220306 RHÔNE F) (23, 24)

WADDINGTON, Prof. D. J., Department of Chemistry, University of York, Heslington, York YO1 5DD, UK
(TEL: +44 (904) 59861. TELEX: 57933 YORKUL G) (11, 13)

WADSÖ, Prof. I., Chemical Center, Thermochemistry, University of Lund, Box 740, S-220 07 Lund 7, Sweden
(TEL: +46 (46) 108188. TELEX: S-33533 LUNIVER) (13, 15)

WAGMAN, Mr. D. D., Blaasstrasse 25/9, A-1190 Wien, Austria
(TEL: +43 (222) 3440404) (15)

WAGNER, Prof. H. Gg., Institut für Physikalische Chemie der Universität, Tammannstrasse 6, D-3400 Göttingen/Weende, Federal Republic of Germany
(TEL: +49 (551) 393111. TELEX: 96703 UNIGOE D) (10)

WAGNER, Dr. M. H., Sigri Elektrographit, Bereich Forschung und Entwicklung, Postfach 1160, D-8901 Meitingen, Federal Republic of Germany (25)

WAGNIÈRE, Prof. G. H., Physikalisch-Chemisches Institut der Universität Zürich, Winterthurerstrasse 190, CH-8057 Zürich, Switzerland
(TEL: +41 (1) 257 44 14. TELEX: 55575 UNIZI CH) (12)

WAKEHAM, Dr. W. A., Department of Chemical Engineering and Chemical Technology, Imperial College of Science and Technology, Prince Consort Road, South Kensington, London SW7 2BY, UK
(TEL: +44 (1) 589 5111. TELEX: 261503 G) (16)

WALDIE, Dr. B., Chemical Engineering Department, Heriot-Watt University, Chambers Street, Edinburgh EH1 1HX, UK
(TEL: +44 (31) 225 8432) (15)

WALKER, Dr. C. W. H., Department of Pathology, McMaster University Medical Centre, Hamilton, Ontario L8S 4J9, Canada
(TEL: +1 (416) 521 2100. TELEX: 0618347 CA) (36)

WALTKING, Mr. A. E., Best Foods Research Center, CPC International Inc., POB 1534, 1120 Commerce Avenue, Union, New Jersey 07083, USA
(TEL: +1 (201) 688 9000) (33)

WANDELBURG, Prof. K., Bundesanstalt für Materialprüfung (BAM), Fachgruppe 1.4 "Anorganisch-chemische Untersuchungen", Unter den Eichen 87, D-1000 Berlin 45, Federal Republic of Germany
(TEL: +49 (30) 81049400. TELEX: 183261 BAMB D) (15)

WANG, Prof. S. M., No. 2, Kwang Chow Street, Taipei, Taiwan
(TEL: 02 3615705. TELEX: 28164 RO CNSC) (11)

WÄNNINEN, Prof. E. V., Department of Analytical Chemistry, Åbo Akademi, Biskopsgatan 8, SF-20500 Åbo 50, Finland
(TEL: 335133. TELEX: 62301 AABIB SF) (27)

WARD, Dr. M. K., Royal Victoria Infirmary, Queen Victoria Road, Newcastle upon Tyne NE1 4LP, UK
(TEL: +44 (632) 325131) (38)

WARD, Mrs. R. J., Division of Clinical Chemistry, Clinical Research Centre, Medical Research Council, Watford Road, Harrow, Middlesex HA1 3UJ, UK
(TEL: +44 (1) 864 5311) (38)

WARNER, Dr. F. P., Polymer Laboratories Ltd., Essex Road, Church Streeton, Shropshire SY6 6A, UK (25)

WATANABE, Prof. K., Department of Mechanical Engineering, Faculty of Science and Technology, Keio University, 3-14-1 Hiyoshi, 3-chome, Kohoku-ku, Yokohama, 223 Japan
(TELEX: 2322198 KOMLIC J) (16)

WATKINSON, Dr. J. H., Ruakura Soil and Plant Research Station, Hamilton, New Zealand (38)

WEISENBERG, Dr. E., Institute for the Standardization and Control of Pharmaceuticals, POB 1457, Jerusalem, Israel
(TEL: +972 (2) 247402. TELEX: 26137 HEAL IL) (38)

WEISS, Prof. A., Institut für Anorganische Chemie der Universität München, Meiserstrasse 1, D-8000 München 2, Federal Republic of Germany (18)

WEISSBACH, Dr. O., Beilstein Institut für Literatur der Organischen Chemie,
Carl Bosch Haus, Varrentrappstrasse 40-42, D-6000 Frankfurt/Main 90,
Federal Republic of Germany
(TEL: +49 (611) 7917252) (21, 22)

WELLS, Dr. D. E., Department of Agriculture and Fisheries for Scotland,
Freshwater Fisheries Laboratory, Faskally, Pitlochry, Perthshire PH16 5LB, UK
(TEL: +44 (796) 2060. TELEX: 73587 MARLAB G) (27, 28)

WELLS, Prof. P. B., Department of Chemistry, University of Hull, Hull HU6 7RX, UK
(TEL: +44 (482) 497660/46311. TELEX: 52530 UNIHUL G) (17, 18)

WELZ, Dr. B., Perkin-Elmer & Co. GmbH, Bodenseewerk, Postfach 1120,
D-7770 Überlingen, Federal Republic of Germany
(TEL: +49 (7551) 813791. TELEX: 733902 D) (38)

WENDORFF, Dr. J., Deutsches Kunststoff-Institut, Schlossgartenstrasse 6 R,
D-6100 Darmstadt, Federal Republic of Germany
(TEL: +49 (6151) 162104) (26)

WESSELS, Prof. H., Bundesanstalt für Fettforschung, Piusallee 68-76,
D-4400 Münster, Federal Republic of Germany
(TEL: +49 (251) 43510) (33)

WEST, Prof. T. S., Macaulay Institute for Soil Research,
Craigiebuckler, Aberdeen AB9 2QJ, UK
(TEL: +44 (224) 38611. TELEX: 73458 UNIABN G) (, 10, 11, 13)

WESTMORELAND, Dr. D. G., Rohm and Haas Co., 727 Norristown Road,
Spring House, Pennsylvania 19477, USA
(TEL: +1 (215) 641 2033. TELEX: 2156417867) (27, 28)

WETTON, Dr. R. E., Department of Chemistry, Loughborough University
of Technology, Loughborough, Leicestershire LE11 3TU, UK
(TEL: +44 (509) 263171. TELEX: 34319 G) (24)

WHIFFEN, Prof. D. H., Department of Physical Chemistry, School of Chemistry, Bedson
Building, University of Newcastle upon Tyne, Newcastle upon Tyne NE1 7RU, UK
(TEL: +44 (632) 328511. TELEX: 53654 UNINEW G) (11, 12, 13, 14, 15)

WHITE, Jr., Dr. H. J., Room A316, Building 221, National Bureau of Standards,
US Department of Commerce, Washington, DC 20234, USA
(TEL: +1 (301) 921 2581. TELEX: 898493 GARG) (16)

WHITE, Prof. J. L., Polymer Engineering Center, University of Akron, Akron,
Ohio 44325, USA
(TEL: +1 (216) 375 7817) (24)

VAN WIJK, Drs. H. F., Hoofdgroep Maatschappelijke Technologie, Organisatie
voor Toegepast Natuurwetenschappelijk Onderzoek, Postbus 108,
3700 AC Zeist, Netherlands
(TEL: +31 (3404) 55444. TELEX: 40022 CIVO NL) (15)

WILHELM, Dr. E., Institut für Physikalische Chemie der Universität Wien,
Währingerstrasse 42, A-1090 Wien, Austria
(TEL: +43 (222) 343616/7) (30)

WILHELMI, Prof. H., Institut für Industrieofenbau und Wärmetechnik im Hüttenwesen
der Rheinisch-Westfälischen Technischen Hochschule Aachen, Kopernikusstrasse 16,
D-5100 Aachen, Federal Republic of Germany
(TEL: +49 (241) 805935. TELEX: 832704 D) (15)

WILKINSON, Prof. F., Department of Chemistry, Loughborough University
of Technology, Loughborough, Leicestershire LE11 3TU, UK
(TEL: +44 (509) 263171. TELEX: 34319 G) (22)

WILLEY, Prof. JOAN D., Department of Chemistry, University of North Carolina,
POB 3725, Wilmington, North Carolina 28406, USA (30)

WILLIAMS, Prof. D. R., Department of Applied Chemistry, University of Wales Institute of Science and Technology, King Edward VII Avenue, Cardiff CF1 3NU, UK
(TEL: +44 (222) 42522) (37)

WILLIAMS, Dr. D. W., Dow Corning Corpn., Midland, Michigan 48640, USA (26)

WILLIAMS, Dr. I. A., Royal Society of Chemistry, Burlington House, Piccadilly, London W1V 0BN, UK
(TEL: +44 (1) 734 9864. TELEX: 268001 G) (11)

WILLIS, Dr. H. A., 4 Sherrardspark Road, Welwyn Garden City, Hertfordshire AL8 7JP, UK
(TEL: +44 (7073) 23965) (17)

WILLS, Dr. M. R., Clinical Laboratories, Department of Pathology, University of Virginia Medical Center, POB 168, Charlottesville, Virginia 22908, USA
(TEL: +1 (804) 924 2553) (38)

WILSON, Dr. H. F., 3 Joining Brook, Spring House, Pennsylvania 19477, USA
(TEL: +1 (215) 628 8399. TELEX: 845247 C/O MRS. DORIS WHELAN) (12)

WINTER, Prof. H. H., Department of Chemical Engineering, University of Massachusetts, Amherst, Massachusetts 01003, USA
(TEL: +1 (413) 545 0922. TELEX: 95 5491) (24)

WIRTH, Dr. H. O., CIBA-GEIGY Marienberg GmbH, Postfach 209, D-6140 Bensheim, Federal Republic of Germany
(TEL: +49 (6254) 79350. TELEX: 0468408 D) (25)

WITENHAFER, Dr. B. F., BP Goodrich Co., Technical Center, POB 122, Avon Lake, Ohio 44012, USA
(TEL: +1 (216) 933 6182) (25)

WITTMER, Dr. P., BASF Aktiengesellschaft, Kunststoff Laboratorium, WKOA-B1, D-6700 Ludwigshafen/Rhein, Federal Republic of Germany
(TEL: +49 (621) 606393. TELEX: 464811 BASF D) (26)

WOLF, Prof. B. A., Institut für Physikalische Chemie, Fachbereich Chemie, Johannes-Gutenberg Universität Mainz, Postfach 3980, Jakob-Welder-Weg 15, D-6500 Mainz, Federal Republic of Germany
(TEL: +49 (6131) 392491. TELEX: 4187476 UNI D) (30)

WOLFF, Dr. J. P., Institut des Corps Gras, 10/A rue de la Paix, F-75002 Paris, France
(TEL: +33 (1) 296 50 29. TELEX: 230905 STABILI PARIS F) (33)

DE WOLFF, Prof. P. M., Laboratorium voor Technische Natuurkunde, Technische Hogeschool Delft, Postbus 5046, 2600 GA Delft, Netherlands (17)

WOLFRAM, Prof. E., Kolloidkémiai és Kolloidtechnológiai Tanszék, Eötvös Loránd Tudományegyetem, POB 328, H-1445 Budapest VIII, Hungary
(TEL: +36 (1) 187 957) (17)

WOOD, Dr. J. A., Department of Chemical Sciences, Polytechnic, Queensgate, Huddersfield HD1 3DH, UK (31)

WOOLLEY, Prof. E. M., Bartlesville Energy Center, POB 1398, Bartlesville, Oklahoma 74003, USA (31)

WORRELL, Prof. W. L., Department of Metallurgy and Materials Science K-1, University of Pennsylvania, Philadelphia, Pennsylvania 19104, USA
(TEL: +1 (215) 898 8592) (20)

WORSAAE, Dr. U., Medicinsk Laboratorium A/S, Adelgade 7, DK-1304 København K, Denmark
(TEL: +45 (1) 14 65 70) (35, 36)

WORTH, Dr. H. G. J., Department of Clinical Chemistry, Kings Mill Hospital, Mansfield Road, Sutton-in-Ashfield, Nottinghamshire NG17 4JL, UK
(TEL: +44 (623) 22515) (36)

WUNDERLICH, Prof. B., Rensselaer Polytechnic Institute, Troy, New York 12181, USA (26)

WUTZEL, Prof. H. J. G., Höhere Bundeslehr- und Versuchsanstalt für Chemische Industrie, Rosensteingasse 79, A-1170 Wien, Austria
(TEL: +43 (222) 461480) (32)

YAGODIN, Prof. G. A., Moscow D I Mendeleev Institute of Chemical Technology, Miusskaya Ploshchad 9, SU-125047 Moscow, A-47, USSR (30)

YALKOWSKY, Prof. S. H., College of Pharmacy, University of Arizona, Tucson, Arizona 85721, USA
(TEL: +1 (602) 626 1289) (30)

YAMADA, Dr. A., Department of Material Science and Technology, Technical University of Nagaoka, Nagaoka, 949-54 Japan
(TEL: +81 (258) 46 6000) (16)

YAMAMURA, Prof. Y., Osaka University, Yamada-oka 1-1, Suita-city, Osaka-fu, 565 Japan
(TEL: +81 (6) 877 5111) (35)

YAMASAKI, Prof. K., 1-28 Zendana-cho, Mizuho, Nagoya, 467 Japan
(TEL: +81 (52) 841 5352) (19)

YAMAUCHI, Prof. O., Faculty of Pharmaceutical Sciences, Kanazawa University, 13-1 Takara-machi, Kanazawa, 920 Japan
(TEL: +81 (762) 62 8151) (29, 30)

YANAGIDA, Prof. H., Department of Industrial Chemistry, Faculty of Engineering, University of Tokyo, 3-1 Hongo 7-chome, Bunkyo-ku, Tokyo, 113 Japan
(TEL: +81 (3) 812 2111. TELEX: 25510 UNITOKYO J) (20)

YANG, Dr. G. Q., Department of Nutrition and Food Hygiene, Institute of Health, Chinese Academy of Medical Sciences, Beijing, China (38)

YARIV, Prof. F., Institute of Chemistry, Hebrew University of Jerusalem, Jerusalem 91904, Israel
(TELEX: 25391 HU IL) (6)

YEN, Prof. T.-S., Chinese Academy of Sciences, San Li-He, Beijing, China
(TEL: 868361. TELEX: 22474 ASCHI CN) (20)

YEOH, Dr. B. G., 22 Jalan SS 4D/12, People's Park, Petaling Jaya, Selangor, Malaysia
(TEL: +60 (3) 362601. TELEX: SIRIM 38672 MA) (33)

YONG, Dr. W. M., Rubber Research Institute of Malaysia, c/o POB 150, Kuala Lumpur, Malaysia
(TEL: +60 (3) 467033. TELEX: 30369 RRIM MA) (27)

YOSHIOKA, Dr. A., Nippon Zeon Co. Ltd., 2-6-1 Marunouchi, Chiyoda-ku, Tokyo, 100 Japan
(TEL: +81 (3) 216 1771. TELEX: 222 4001 NIPOZE J) (24)

YOUNG, Dr. C. L., Department of Physical Chemistry, University of Melbourne, Parkville, Victoria 3052, Australia
(TEL: +61 (3) 341 6465. TELEX: AA35185 UNIMEL) (30)

YOUNG, Dr. D. S., Department of Laboratory Medicine, 210 Hilton Building, Mayo Clinic, Rochester, Minnesota 55905, USA
(TEL: +1 (507) 284 3116) (35)

YOUNG, Mr. J., Borg-Warner Chemicals, PO Box 8122, Amsterdam, Netherlands
(TEL: +31 (20) 139833. TELEX: 16257 NL) (24)

ZACHARIASEN, Dr. H., Falconbridge Nikkelverk Aktieselskap, POB 457, N-4601 Kristiansand S, Norway
(TEL: +47 (42) 25500) (37)

ZACHMANN, Prof. H. G., Institut für Anorganische und Angewandte Chemie, Martin-Luther-King-Platz 6, D-2000 Hamburg 13, Federal Republic of Germany
(TEL: +49 (40) 893544) (24)

ZAIKOV, Prof. G. E., Institute of Chemical Physics, Academy of Sciences of USSR, Kosygin Street 4, SU-117977, SZD-1, Moscow, V-334, USSR
(TEL: +7 (095) 139 7191) (10, 12)

ZAMARAEV, Prof. K. I., Institute of Catalysis, 5 Prospekt Academica Lavryentyeva, SU-630090 Novosibirsk 90, USSR
(TEL: +7 (3832) 655 756) (15)

ZANDER, Prof. M., Rütgerswerke AG, Postfach 504, Kekuléstrasse 30, D-4620 Castrop-Rauxel, Federal Republic of Germany
(TEL: +49 (2305) 705337. TELEX: 08229517 D) (28)

ZEMBURA, Prof. Z., ul. Goetla 10, PL-30 065 Kraków, Poland (16)

ZENDER, Dr. R., Laboratoire, Hôpital de la Chaux-de-Fonds, CH-2301 La Chaux-de-Fonds, Switzerland
(TEL: +41 (39) 21 11 91) (12, 36)

ZERBI, Prof. G., Dipartimento di Chimica Industriale e Ingegneria Chimica, Politecnico di Milano, Piazza Leonardo da Vinci 32, I-20133 Milano, Italy
(TEL: +39 (2) 292105/292125-6. TELEX: 333467 POLIMI I) (17)

ZEWAIL, Prof. A. H., Department of Chemistry, Mail Code 127-72, California Institute of Technology, Pasadena, California 91125, USA
(TEL: +1 (213) 356 6536. TELEX: 675425) (17)

ZIELENKIEWICZ, Prof. W., Instytut Chemii Fizycznej, Polskiej Akademii Nauk, ul. Kasprzaka 44-52, PL-01 224 Warszawa, Poland
(TEL: +48 (22) 327343. TELEX: 817097 ICHF PL) (16)

ZIKMUND, Dr. M., Ústav Anorganickej Chemie, Slovenská Akadémia Vĕd, Dúbravská cesta, CS-809 34 Bratislava, Czechoslovakia
(TEL: +42 (7) 45571) (19)

ZIMMER, Prof. K., Szervetlen és Analitikai Kémiai Tanszék, Eötvös Loránd Tudományegyetem, POB 123, H-1443 Budapest, Hungary
(TEL: +36 (1) 139 658) (28)

ZIMMERMANN, Prof. H., Institut für Polymerenchemie "Erich Correns", Akademie der Wissenschaften der DDR, Kantstrasse 55, DDR-1530 Teltow-Seehof, German Democratic Republic
(TEL: 4831. TELEX: 15322 DD) (24)

ZINDER, Dr. O., Clinical Biochemistry Department, Rambam Medical Centre, Haifa, Israel
(TEL: +972 (4) 533111) (36)

ZOLOTOV, Prof. Yu. A., V I Vernadskii Institute of Geochemistry and Analytical Chemistry, Academy of Sciences of USSR, Kosygin Street 19, SU-117975, SZD-1, Moscow, V-334, USSR
(TEL: +7 (095) 139 7041) (12, 27)

ZORLL, Dr. U., Forschungsinstitut für Pigmente und Lacke eV, Wiederholdstrasse 10/1, D-7000 Stuttgart 1, Federal Republic of Germany
(TEL: +49 (711) 297549) (25)

ZVONAR, Mr. V., Výzkumny Ústav Synthetických Pryskyřic a Laků, S K Neumanna 1316, Pardubice, Czechoslovakia (25)

ZWENNIS, Dr. W. C. M., TNO Medisch Biologisch Laboratorium, Postbus 45, 2280 AA Rijswijk, Netherlands
(TEL: +31 (15) 138777. TELEX: 38034 PMTNO NL) (38)

ZWERENZ, Dr. G., Österreichische Unilever GmbH, Breitenfurterstrasse 239, A-1230 Wien, Austria
(TEL: +43 (222) 842515/862445) (33)

ZWIERZYKOWSKI, Prof. W., Instytut Chemii i Technologii Organicznej oraz Żywnosciowej, Politechnika Gdańska, ul. Majakowskiego 11-12, PL-80 952 Gdańsk-Wrzeszcz, Poland (33)

AMENDMENTS

(To be published from time to time in CI)

CURRENT PROGRAMMES 1984

The listing presents brief details of all official programmes of the Union in operation in its special Committees and in Commissions of the Physical Chemistry, Inorganic Chemistry, Organic Chemistry, Macromolecular, Analytical Chemistry, Applied Chemistry, and Clinical Chemistry Divisions. The name of each project coordinator is given and the dates for starting and for completion or review. There are also brief comments on some of the programmes, indicating joint interests with other IUPAC bodies or a general interest in broad areas not otherwise identified, e.g., nomenclature, environment, health.

The first listing was produced in June 1976; the second, third, fourth and fifth appeared in May of 1977, 1979, 1981, and 1983; and this is the sixth edition. The present intention is that the next issue will not be published before 1986. Further details of individual programmes are available from the Secretary of the relevant Division or from the Secretary of the appropriate Commission or special Committee.

Appendix I (page 171) presents a listing of reports of IUPAC activities published during 1983, and Appendix II (page 175) lists forthcoming IUPAC-sponsored symposia.

GUIDELINES FOR IUPAC PROJECTS

1. General Criteria

IUPAC projects should satisfy at least one of the following key criteria:

1.1 they should be related to the needs of the chemists in the world, not just in a country or a region,

1.2 they should be related to the role of chemistry for the needs of mankind,

1.3 they should best be approached by an international team such as IUPAC can assemble.

2. Examples

Thus, IUPAC projects can comprise:

2.1 Studies of problems of international nomenclature, symbols, terminology and conventions, as the need develops; this includes permanent activities (e.g. nomenclature of inorganic chemistry), or ad hoc studies (e.g. symbols and units for biotechnology, nomenclature of synthetic and natural zeolites).

2.2 Large future-oriented projects important for the position of chemistry in the world or for the needs of mankind (CHEMRAWN, teaching of chemistry, problems of publications or of computers, etc.).

2.3 Compilations of data best undertaken by an international team, for international use (e.g. solubility data).

2.4 Unification of "approved" experimental methods (in particular related to analytical problems, characterization of materials, etc.), in domains where the specialized organizations (ISO, EEC, NBS) are not active, or in close cooperation with them, preferably as contract work.

3. Additional Comments

3.1 The scientific or industrial importance of a field in chemistry is not in itself sufficient justification for its presence in IUPAC as a Division or Commission, although such relevance must be considered a criterion for continuing ongoing activities as well as for establishing new programmes.

3.2 The organization of symposia is not by itself justification for the existence of a Commission: it is part of the normal activity of the Divisions or of the group of Division Presidents, of course helped by suggestions from members of Commissions, Subcommittees and Bureau. Such IUPAC bodies are important as an International Advisory Committee for establishment of the scientific programme; all other organizational arrangements have to be made by a local (or national) committee.

3.3 An IUPAC project should be clearly defined as continuing or limited in time (which is the normal case), with initiation and tentative final dates, with a coordinator and a "budget" (depending on the case, in money, or/and in terms of number of meetings, or other investments of effort).

CURRENT PROGRAMMES - PROJECT NUMBERING

In Current Programmes 1983 the numbering of new projects, started since the 1981 issue, continued consecutively from the last number used in the 1981 issue. In the present edition the numbering has again been continued consecutively, this time from the 1983 issue, and those projects approved since the General Assembly in 1983 have been included under the heading "Project(s) approved since August 1983". Projects which have been abandoned or terminated without publication have been identified.

COMPLETION OR *REVIEW DATE OF PROJECT

NPR = No progress reported.

REVISED PROCEDURE FOR COMMENT AND APPROVAL OF IUPAC RECOMMEMDATIONS ON NOMENCLATURE AND SYMBOLS

At Leuven in 1981, at the instigation of Dr. K. L. Loening (Chairman of IDCNS), the IUPAC Executive Committee approved in principle a new system which avoided the double publication of recommendations (in provisional and then revised form) in PAC, but still provided the opportunity for worldwide comment on these provisional proposals. However, this system was not put into effect because of reorganization of the IUPAC Secretariat which, under the system, would have been heavily involved in distribution of provisional proposals. The 'Leuven' procedure has now itself been revised and its successor is to run for a two-year trial period between General Assemblies 1983 and 1985.

1. Provisional recommendations by a Commission should be submitted to the Secretariat via, and with the approval of, the Commission Chairman and the Division President. There should normally have been taken into account comments from about fifteen experts on the subject matter of the document, including about three manuscript editors of journals that publish substantial numbers of papers in the relevant field of chemistry. The names and addresses of the experts and/or editors that have been consulted must be submitted with the provisional recommendations; Titular or Associate Members of the Commission in question should only be included on the list if they have specific expertise in the subject.

2. The provisional recommendations sent to the Secretariat must include:

   (a) a 200-word synopsis;

   (b) the name and address of the Commission Member acting as document editor to whom comments should be sent;

   (c) the name of the Commission Member who will eventually prepare camera-ready copy for publication in PAC.

3. The provisional recommendations will be sent from the Secretariat to the Chairman and Secretary of IDCNS for approval, with copy letters to Members of IDCNS (giving them the opportunity to request a review copy for comment within three months to IDCNS Officers). The names and addresses of the experts and/or editors consulted by the relevant Commissions (see 1 above) will also be sent to the IDCNS Officers to ensure that they do not approach the same persons (see 4).

4. The IDCNS Chairman, or by designation the Secretary, on suggestions from Members of IDCNS and/or if he considers it to be desirable, may instruct the Secretariat to send the provisional recommendations to further experts and/or editors whose names and address are supplied to him. The IDCNS Chairman or Secretary will also signify his approval, or otherwise, of the synopsis and of public distribution of the provisional recommendations.

5. The synopsis, if approved, will be sent from the Secretariat, with a note that a copy of the provisional recommendations may be obtained <u>only</u> for comment from various national/regional centres, for publication in:

   (a) <u>Chemistry International</u> (CI);

   (b) national chemistry news journals via certain national/regional centres.

6. All those who are sent, or send for, a copy of the provisional recommendations will be asked to return their comments to the document editor within eight months of the approval of IDCNS.

7. Not less than eight months after approval by IDCNS the provisional recommendations will be revised, as necessary, by the Commission to take into account comments received.

8. The revised recommendations will be resubmitted to the Secretariat, via the Commission Chairman and Division President, for approval by IDCNS.

9. The revised recommendations, with their date (year) of adoption, will be published in PAC.

10. Publication of the revised recommendations will be announced in CI.

NATIONAL/REGIONAL CENTRES

The following organizations have agreed to act as national/regional centres under these new arrangements:

Australia and New Zealand

Mr. R. Schoenfeld, Australian Journal of Chemistry, POB 89, 314 Albert Street, East Melbourne, Victoria 3002, Australia.

Canada

Prof. J. ApSimon, Department of Chemistry, Carleton University, Ottawa, Ontario K1S 5B6, Canada.

China

Prof. Cui Meng-Yuan, Chinese Chemical Society, POB 2709, Beijing, China.

Prof. F. C. Chen, Department of Chemistry, National Taiwan University, POB 30-373, Taipei 107, Taiwan.

Eastern Europe (Bulgaria, Czechoslovakia, German Democratic Republic, Hungary, Poland, Romania)

Prof. E. Pungor, Altalános és Analitikai Kémiai Tanszék, Budapesti Müszaki Egyetem, Gellért tér 4, H-1521 Budapest XI, Hungary.

France

Prof. Y. Jeannin, Comité National de la Chimie, 28 rue Saint-Dominique, F-75007 Paris, France.

FRG, Austria and Switzerland

Dr. W. Fritsche, Deutscher Zentralausschuss für Chemie, Postfach 900440, Varrentrappstrasse 40-42, D-6000 Frankfurt/Main 90, Federal Republic of Germany.

India

Mr. A. K. Bose, Indian National Science Academy, Bahadur Shah Zafar Marg, New Delhi-110002, India.

Israel

Prof. S. Sarel, Department of Pharmaceutical Chemistry, School of Pharmacy, Hebrew University of Jerusalem, POB 12065, Jerusalem, Israel.

Japan

Dr. H. Hamada, Executive Director, Chemical Society of Japan, 1-5, Kanda Surugadai, Chiyoda-ku, Tokyo, 101 Japan.

Netherlands

Ir. E. J. de Ryck van der Gracht, Koninklijke Nederlandse Chemische Vereniging, Postbus 90613, 2509 LP Den Haag, Netherlands.

Scandinavia

Dr. K. Samuelsson, Acta Chemica Scandinavica, Wallingatan 26 A, S-111 24 Stockholm, Sweden.

Republic of South Africa

Dr. J. R. Bull, Organic Chemistry Division, National Chemical Research Laboratory, POB 395, Pretoria 0001, Republic of South Africa.

UK

Dr. I. A. Williams, Royal Society of Chemistry, Burlington House, Piccadilly, London W1V 0BN, UK.

USA

Dr. Marianne Brogan, Journals Department, American Chemical Society, POB 3330, 2540 Olentangy River Road, Columbus, Ohio 43210, USA.

USSR

Prof. G. B. Bokij, Commission on Nomenclature, National Committee of Soviet Chemists, Academy of Sciences of USSR, Kosygin Street 4, SU-117977 Moscow V-334, USSR.

Negotiations are in an advanced state for a number of other centres and these will be announced in CI as the details are finalized. Meanwhile, copies of provisional recommendations may be obtained from the national/regional centre nearest to you or as a last resort from:

Dr. K. L. Loening, Chemical Abstracts Service, POB 3012, 2540 Olentangy River Road, Columbus, Ohio 43210, USA.

Copies cannot be obtained from the IUPAC Secretariat.

REPUBLICATION/TRANSLATION OF IUPAC COMMISSION REPORTS

The Union intends to introduce in 1985 a generous and simple policy for reproduction of material from its publications, especially with regard to reports on nomenclature/symbols and other technical matters published in PAC and material published in CI.

Look out for this announcement in PAC and CI towards the end of 1984/early in 1985.

LIST OF ABBREVIATIONS

ACS	-	American Chemical Society
ASTM	-	American Society for Testing and Materials
BIPM	-	Bureau International des Poids et Mesures
CAC	-	Codex Alimentarius Commission
CHEMRAWN	-	Chemical Research Applied to World Needs
CI	-	Chemistry International, news magazine of IUPAC
CIPM	-	Comité International des Poids et Mesures
COCI	-	IUPAC Committee on Chemistry and Industry
CODATA	-	ICSU Committee on Data for Science and Technology
CODEX	-	Joint FAO/WHO Committee on Pesticide Residues
COSTED	-	ICSU Committee on Science and Technology in Developing Countries
CTC	-	IUPAC Committee on Teaching of Chemistry
CTS	-	ICSU Committee on the Teaching of Science
EEC	-	European Economic Community
FAO	-	United Nations Food and Agriculture Organization
IAEA	-	International Atomic Energy Agency
IARC	-	International Agency for Research on Cancer
ICSH	-	International Committee for Standardization in Haematology
ICSU	-	International Council of Scientific Unions
IDCNS	-	Interdivisional Committee on Nomenclature and Symbols
IDF	-	International Dairy Federation
IFCC	-	International Federation of Clinical Chemistry
ILO	-	United Nations International Labour Organization
IOBC	-	International Organization for Biological Control
IOCD	-	International Organization for Chemical Sciences in Development
ISO	-	International Standards Organization
ISO/TC 12	-	ISO Technical Committee on Quantities, Units, Symbols, Conversion Factors and Conversion Tables
ISO/TC 156	-	ISO Technical Committee on Corrosion of Metals and Alloys
IUB	-	International Union of Biochemistry
IUCr	-	International Union of Crystallography
IUCS	-	Inter-Union Commission on Spectroscopy

IUPAB	-	International Union of Pure and Applied Biophysics
IUPAP	-	International Union of Pure and Applied Physics
JAOAC	-	Journal of Association of Official Analytical Chemists
JCBN	-	IUB-IUPAC Joint Commission on Biochemical Nomenclature
JMPR	-	Joint Meeting of FAO/WHO Experts on Pesticide Residues
NATO	-	North Atlantic Treaty Organization
NBS	-	US National Bureau of Standards
OICC	-	Office Internationale du Cacao et Chocolat
OIML	-	Organisation Internationale de Métrologie Légale
PAC	-	Pure and Applied Chemistry, official journal of IUPAC
REMCO	-	ISO Committee on Reference Materials
SCAR	-	Scientific Committee on Antartic Research
SCOPE	-	Scientific Committee on Problems of the Environment
SCOR	-	Scientific Committee on Oceanic Research
SUN	-	IUPAP Commission on Symbols, Units, Nomenclature, Atomic Masses, and Fundamental Constants
UNESCO	-	United Nations Educational, Scientific and Cultural Organization
WHO	-	United Nations World Health Organization

STANDING COMMITTEES

CHEMRAWN COMMITTEE

PROJECT TITLE		COORDINATOR	STARTING DATE	COMPLETION OR * REVIEW DATE	OTHER INTERESTS AND COMMENTS
	Project 5/81 withdrawn				
2/81	CHEMRAWN III World Conference on Resource Material Conversion : (Bio) Chemical Process Bridges to meet Future Needs (The Hague, 84.06.25-29)	J.B. Donnet	1980	1984	COCI
3/81	CHEMRAWN IV Chemical Resources of the Global Oceans (Woods Hole, Massachusetts, 85.09.22-28)	C.G. Overberger	1979	1985	SCOPE, SCOR, SCAR
4/81	CHEMRAWN V Contribution of Chemistry to Human Health : the New Frontiers (Revised title) (Federal Republic of Germany, 1986)	H. Egan	1980	1986	IARC, IUB
	- - - - - - - - - - - - Project approved from August 1983 - - - - - - - - - - - -				
6/83	CHEMRAWN VI Special Materials to Meet New Needs : Energy, Transportation and Communication (Japan, 1987)	M. Uchida	1983	1987	Japan CHEMRAWN Committee, COCI

COMMITTEE ON TEACHING OF CHEMISTRY (CTC)

	PROJECT TITLE	COORDINATOR	STARTING DATE	COMPLETION OR * REVIEW DATE	OTHER INTERESTS AND COMMENTS
	Projects 6/81, 9-13/83 have been completed				
8/81	UNESCO Sourcebook on Chemical Education in Secondary Schools : 'Teaching School Chemistry'	D.J. Waddington	1978	1984	UNESCO
14/83	Booklet on Safety in School Laboratories	T.R. Hitchings	1982	1985	
15/83	Collaboration in Series of Workshops for Science Technicians	D.J. Waddington	1982	Ongoing	ICSU, CTS, UNESCO
16/83	Collaboration with Conference on Science for Development (Bangalore, August 1985)	D.J. Waddington	1982	1985	ICSU, CTS
17/83	International Newsletter on Chemical Education	D.J. Waddington	1975	Ongoing	UNESCO
	- - - - - - Projects approved from August 1983 - - - - - -				
18/83	Second Booklet on Locally-Produced Equipment for Chemistry Teaching	K.V. Sane	1983	1984	UNESCO
19/83	Proceedings of Third Regional Workshop on Low-Cost Locally-Produced Equipment (Copenhagen, July 1983)	E.W. Thulstrup	1983	1984	UNESCO, ICSU, CTS, Danish Ministry of Education
20/83	Fourth Regional Workshop on Locally-Produced Equipment for Chemistry Teaching (Bangladesh, 1984)	S.Z. Haider K.V. Sane	1983	1984	UNESCO, Commonwealth Foundation, ICSU, COSTED, CTS

COMMITTEE ON TEACHING OF CHEMISTRY (CTC) (CONTINUED)

PROJECT	TITLE	COORDINATOR	STARTING DATE	COMPLETION OR * REVIEW DATE	OTHER INTERESTS AND COMMENTS
21/83	Fifth Regional Workshop on Locally-Produced Equipment for Chemistry Teaching (Sri Lanka, 1985)	R.S. Ramakrishna K.V. Sane	1983	1985	UNESCO, COSTED, Commonwealth Foundation, ICSU, CTS
22/83	Seminar on Locally-Produced Equipment for Chemistry Teaching (Yugoslavia, 1985)	A. Kornhauser K.V. Sane	1983	1985	IOCD, UNESCO

IUB-IUPAC JOINT COMMISSION ON BIOCHEMICAL NOMENCLATURE (JCBN)

PROJECT	TITLE	COORDINATOR	STARTING DATE	COMPLETION OR * REVIEW DATE	OTHER INTERESTS AND COMMENTS
5/81	Prostaglandins and Thromboxanes (Revised title) - Provisional	K.L. Loening	1978	* 1984	
6/81	Amino Acids and Derivatives - Revised	H.B.F. Dixon	1979	1984	
7/81	Amine Neurotransmitters and their Analogues (Revised title) - Provisional	K.L. Loening	1978	1984	
10/81	Carbohydrate Nomenclature - Revised	D. Horton	1981	* 1985	
11/81	Glycoproteins - Provisional	N. Sharon	1979	1984	
13/81	Folic Acid - Revised	R. Blakley	1981	* 1984	

IUB-IUPAB-IUPAC COMMISSION ON BIOTHERMODYNAMICS (CB)

PROJECT	TITLE	COORDINATOR	STARTING DATE	COMPLETION OR * REVIEW DATE	OTHER INTERESTS AND COMMENTS
1/81	Compilation of Thermodynamic Data of Biological Interest	H.-J. Hinz	1980	1984	CODATA, IUB, IUPAB
2/81	Measurements and Presentation of Biothermodynamic Data - Provisional	H. Eisenberg	1977	1984	IUB, IUPAB
5/81	Guidelines for Characterization of Calorimetric Instruments (Revised title)	I. Wadsö	1979	1985	IUB, IUPAB

PHYSICAL CHEMISTRY DIVISION (I)

President: Dr. D. R. Lide, Jr. (Office of Standard Reference Data, A323 Physics Building, National Bureau of Standards, US Department of Commerce, Washington, DC 20234, USA)

Vice-President: Prof. K. I. Zamaraev

Secretary: Dr. E. D. Becker

SUBCOMMITTEE ON PHYSICOCHEMICAL MEASUREMENTS AND STANDARDS

	PROJECT TITLE	COORDINATOR	STARTING DATE	COMPLETION OR * REVIEW DATE	OTHER INTERESTS AND COMMENTS
	Projects 2/81 Bibliography published (see Appendix I), 4 and 5/81 included in 8/83, 10/83 abandoned				
8/83	Recommended Reference Materials and Methods for the Realization of Physicochemical Properties (Book Project)	K.N. Marsh	1981	1985	Commissions I.2, I.3, I.5, V.5
9/83	Use of Terms in Describing the Uncertainty of a Measurement (with particular emphasis on Reference Materials)	S.D. Rasberry	1981	* 1985	Commission I.2, Divisions V, VI, VII, BIPM, OIML, ISO

SUBCOMMITTEE ON PLASMA CHEMISTRY

	PROJECT TITLE	COORDINATOR	STARTING DATE	COMPLETION OR * REVIEW DATE	OTHER INTERESTS AND COMMENTS
	Projects 5.1/83 and 5.2/83 were published in 1982 and 1983 respectively under the auspices of Tubingen University				
1/81	Solid Particles Velocity Measurements in Thermal Plasmas (Revised title)	P. Fauchais	1979	* 1985	
7/83	Mass Spectrometric Diagnostic in Plasma Chemistry	W. Lindinger	1980	NPR	

COMMISSION ON PHYSICOCHEMICAL SYMBOLS, TERMINOLOGY, AND UNITS (I.1)

PROJECT	TITLE	COORDINATOR	STARTING DATE	COMPLETION OR * REVIEW DATE	OTHER INTERESTS AND COMMENTS
2/81	Coordination of Appendices to Manual of Symbols and Terminology	K. Kuchitsu	Before 1976	Ongoing	Commission I.2, IDCNS
3/81	New Edition of Manual of Symbols and Terminology	K. Kuchitsu	1981	1985	Commissions I.2, I.6, ISO/TC 12, IUPAP/SUN, CIPM
5/83	Promotion of the Republications and Translations of the IUPAC Manual	K. Kuchitsu	1981	Ongoing	ISO/TC 12, IUPAP/SUN, CIPM
6/83	Edition of Abbreviated List of Symbols, Terminology and Units	A. Schuyff	1981	1985	All Commissions in Division I, CTC, ISO/TC 12, IUPAP/SUN, CIPM

COMMISSION ON THERMODYNAMICS (I.2)

PROJECT	TITLE	COORDINATOR	STARTING DATE	COMPLETION OR * REVIEW DATE	OTHER INTERESTS AND COMMENTS
	Project 2/81 transferred to CODATA				
5/83	Revision of the Manual and its Appendices on Topics Related to Thermodynamics	G. Somsen	1982	1985	
	- - - - - Projects approved from August 1983 - - - - -				
6/83	Standard States of Mixtures	M.T. Rätzsch	1983	* 1985	
7/83	Symbols and Nomenclature in Thermodynamics	R.D. Freeman	1983	* 1985	

SUBCOMMITTEE ON THERMODYNAMIC TABLES

PROJECT	TITLE	COORDINATOR	STARTING DATE	COMPLETION OR * REVIEW DATE	OTHER INTERESTS AND COMMENTS
	International Thermodynamic Tables of the Fluid State				Subcommittee on Physicochemical Measurements and Standards, COCI
1/81	Vol.8 : Chlorine	S. Angus	1979	1984	

SUBCOMMITTEE ON THERMODYNAMIC TABLES (CONTINUED)

	PROJECT TITLE	COORDINATOR	STARTING DATE	COMPLETION OR * REVIEW DATE	OTHER INTERESTS AND COMMENTS
	International Thermodynamic Tables of the Fluid State				Subcommittee on Physico-chemical Measurements and Standards, COCI
2/81	Vol.9 : Fluorine	S. Angus	1979	1984	
	Phase Equilibrium and Related Data for Selected Fluids and Fluid Mixtures (FLUIDATA)				
4/81	Enthalpies of Vaporization	H.V. Kehiaian	1981	1985	
	- - - - - - Projects approved from August 1983 - - - - - -				
	International Thermodynamic Tables of the Fluid State				
5/83	Vol.10 : Oxygen	S. Angus	1983	1985	
	Phase Equilibrium and Related Data for Selected Fluids and Fluid Mixtures (FLUIDATA)				
6/83	Liquid-Vapour Equilibrium of n-Alkane +n-Alkanol Systems	H.V. Kehiaian	1983	* 1985	

SUBCOMMITTEE ON TRANSPORT PROPERTIES

1/83	Transport Properties	J. Kestin	1981	* 1985	

COMMISSION ON ELECTROCHEMISTRY (I.3)

Projects 2/81 Provisional Report published (see Appendix I) under the title 'Interphases in Systems of Conducting Phases', 12/83 Commission's work completed, 15/83 Withdrawn pending completion of a feasibility study

4/81	Oxidation-Reduction Potentials in Aqueous Solutions - A Selective and Critical Source Book	A.J. Bard R. Parsons	1977	1984	Joint with Commission V.5

COMMISSION ON ELECTROCHEMISTRY (I.3) (CONTINUED)

PROJECT	TITLE	COORDINATOR	STARTING DATE	COMPLETION OR * REVIEW DATE	OTHER INTERESTS AND COMMENTS
5/81	Conductance Data Compilation	J.C. Justice	Before 1976	Ongoing	
6/81	Kinetic Data Compilation	K. Niki	1977	Ongoing	
7/81	Recommendations on Reporting Electrode Potentials in Non-Aqueous Solvents - Revised	J. Kůta G. Gritzner	1977	1984	
8/81	Fundamentals of Energy Conversion 2. Semiconductor Electrochemistry (Revised title)	R. Memming A.J. Bard B. Miller	1981	1985	
9/81	Nomenclature for Non-Isothermal Systems	J.N. Agar	1981	* 1985	
11/83	Recommendations for Reference Electrode in Molten Salts	J.A. Plambeck A. Yamada K. Niki	1979	* 1985	Commission V.5
13/83	Recommendations for the Analysis of Conductance Data	J.C. Justice	1982	* 1984	
14/83	Fundamentals of Energy Conversion 1. Interfacial Ionic Transfer	K.E. Heusler	1981	1985	
16/83	Corrosion Nomenclature	K.E. Heusler D. Landolt S. Trasatti	1982	* 1984	ISO/TC 156
17/83	Absolute Electrode Potential. An Explanatory Note (Revised title)	S. Trasatti	1982	1984	Commission V.5
	- - - - - - Projects approved from August 1983 - - - - - -				
18/83	Electrode Potentials in Non-Aqueous Solutions (Part I)	G. Gritzner	1982	* 1985	

COMMISSION ON ELECTROCHEMISTRY (I.3) (CONTINUED)

PROJECT TITLE		COORDINATOR	STARTING DATE	COMPLETION OR * REVIEW DATE	OTHER INTERESTS AND COMMENTS
19/83	Terminology for Microelectronic Ion-Sensitive Devices in Electrochemistry	A.K. Covington	1983	* 1985	Joint with Commission V.5

COMMISSION ON CHEMICAL KINETICS (I.4) (ESTABLISHED 1983)

COMMISSION ON MOLECULAR STRUCTURE AND SPECTROSCOPY (I.5)

PROJECT TITLE		COORDINATOR	STARTING DATE	COMPLETION OR * REVIEW DATE	OTHER INTERESTS AND COMMENTS
1/81	Symbols, Terminology and Applications for Electron Spectroscopies	D.W. Turner	1979	1984	Commission I.6, IUCS, IUPAP
7/83	Recommendation for Creating Computerized Data Banks on Molecular Spectroscopy	V. Koptyug	1981	1984	CODATA
8/83	Definitions and Quantities in Optical Spectroscopy	N. Sheppard H.A. Willis	1981	1984	Commissions V.4, VII.2
	- - - - - - - - - - Projects approved from August 1983 - - - - - - - - - -				
9/83	Update of Document for Presentation of Infrared Data	J. Grasselli	1983	1985	
10/83	Secondary Standards for the Determination of Intensities	R.N. Jones	1983	1985	
11/83	Symbols, Quantities, Conventions and Terminology in ESR Spectroscopy	H. Kon	1983	1985	Physical Chemistry Division's Commissions, Division II

SUBCOMMITTEE ON INFRARED AND RAMAN SPECTROSCOPY

	PROJECT TITLE	COORDINATOR	STARTING DATE	COMPLETION OR * REVIEW DATE	OTHER INTERESTS AND COMMENTS
1/81	Specifications for Infrared Reference Spectra of Molecules in Vapour Phase (Revised title)	R.N. Jones J.R. Durig	1979	1984	
2/81	Recommendations for Symbols and Nomenclature in Raman Spectroscopy	J.R. Durig	1977	1985	
3/81	Test Data for Normal Coordinate Calculations	M. Tasumi	1979	* 1985	
4/81	Recommendations for the Presentation of Molecular Parameters Values for Infrared and Raman Intensities Derived from Experimental or from ab initio Calculations (Revised title) (Omitted erroneously from 1983 Current Progammes)	G. Zerbi W.B. Person	1979	1984	
5/81	Tables of Wavenumbers	A.R.H. Cole	1980	Ongoing	
6/83	Glossary of Terms and Usage in Laser Spectroscopy Applied to Chemistry	G. Zerbi	1981	* 1985	Commission III.3

SUBCOMMITTEE ON MASS SPECTROSCOPY

	PROJECT TITLE	COORDINATOR	STARTING DATE	COMPLETION OR * REVIEW DATE	OTHER INTERESTS AND COMMENTS
1/83	Quality Standards for Mass Spectroscopy	J.F.J. Todd	1982	1985	Cooperation with Joint Committee on Atomic and Molecular Physical Data
	- - - - - - - - Project approved from August 1983 - - - - - - - -				
2/83	Recommendation for Nomenclature in Secondary Ion Mass Spectroscopy	J.F.J. Todd	1983	1985	Commission I.6, American Society for Mass Spectrometry, IUPAP

SUBCOMMITTEE ON CHIROPTICAL PHENOMENA (ORD-CD) (DISBANDED AUGUST 1983)

PROJECT	TITLE	COORDINATOR	STARTING DATE	COMPLETION OR * REVIEW DATE	OTHER INTERESTS AND COMMENTS
1/81	Symbolism and Conventions in Circular Dichroism and Optical Rotatory Dispersion (Report prepared for publication)	G. Snatzke	1976	1984	Commission V.3

SUBCOMMITTEE ON NOTATIONS AND CONVENTIONS FOR MOLECULAR SPECTROSCOPY

PROJECT	TITLE	COORDINATOR	STARTING DATE	COMPLETION OR * REVIEW DATE	OTHER INTERESTS AND COMMENTS
1/81	Recommendations for Conventions and Notation for Molecular Spectroscopy	C.J.H. Schutte	1979	1985	ICSU, IUPAP

COMMISSION ON COLLOID AND SURFACE CHEMISTRY INCLUDING CATALYSIS (I.6)

Projects 7/81 Provisional report published 1982, under revised title 'Reporting Experimental Pressure - Area Data with Film Balances', and 8/81 Provisional report published 1983 (see Appendix I)

PROJECT	TITLE	COORDINATOR	STARTING DATE	COMPLETION OR * REVIEW DATE	OTHER INTERESTS AND COMMENTS
2/81	A Lexicon of Experimental Techniques in Surface Chemical Physics (Revised title)	R.S. Hansen	1976	1985	Commissions I.3, I.5.
4/81	Reporting Experimental Data for Solid-Gas Adsorption	K.S.W. Sing	1976	1985	
5/81	Reporting Data on Adsorption from Solution at the Solid/Liquid Interface (Revised title)	D.H. Everett	1976	1985	
9/83	Catalyst Characterization and Standardization	J. Haber	1981	1987	
10/83	Reporting Experimental Pressure - Area Data with Film Balances (Revised title) - Revised	L. Ter-Minassian-Saraga	1982	1984	
	- - - - - - Project approved from August 1983 - - - - - -				
11/83	Terminology and Symbols of Light Scattering - Revised	M. Kerker	1983	1984	

INORGANIC CHEMISTRY DIVISION (II)

President: Dr. Mary L. Good (UOP Corporate Research Center, UOP Inc., Ten UOP Plaza, Des Plaines, Illinois 60016, USA)

Vice-President: Prof. Kazuo Saito

Secretary: Prof. E. Fluck

COMMISSION ON ATOMIC WEIGHTS AND ISOTOPIC ABUNDANCES (II.1)

	PROJECT TITLE	COORDINATOR	STARTING DATE	COMPLETION OR * REVIEW DATE	OTHER INTERESTS AND COMMENTS
	Project 8/83 completed, taken over by Committee on Teaching of Chemistry for publication				
2/81	Review of Atomic Weights and Issue of Tables of Revised Values	R.L. Martin	1902	Ongoing	
3/81	Review of Isotopic Compositions and Issue of a Table of Revision Values	I.L. Barnes	1974	Ongoing	Division II, IUPAC Commissions concerned with isotopic composition of elements found in nature
5/81	Evaluation of the Isotopic Compositions of Non-Terrestrial Materials	W.H. Johnson	1977	* 1985	
9/83	Element by Element Review (of changes which have been made since 1961 to Atomic Weight Values based on assessment of published physical and chemical data)	H.S. Peiser	1981	1984	
10/83	Feasibility Study on Uncertainties	I.L. Barnes	1983	1985	

COMMISSION ON NOMENCLATURE OF INORGANIC CHEMISTRY (II.2)

PROJECT	TITLE	COORDINATOR	STARTING DATE	COMPLETION OR * REVIEW DATE	OTHER INTERESTS AND COMMENTS
	Project 14/81 published (see Appendix 1983 Current Programmes, Page 45, under Commission on Macromolecular Nomenclature)				
	Chapters for Revision of 'Nomenclature of Inorganic Chemistry' (1970 edition : The Red Book)				
1/81	Aims and Functions of Inorganic Nomenclature	T.D. Coyle	1980	1985	Commission III.1
2/81	Grammar of Inorganic Nomenclature	Y. Jeannin	1980	1985	Commission III.1
3/81	Elements, Atoms, and Groups of Atoms (Revised title)	G.J. Leigh	1980	* 1984	Commission II.1
4/81	Formulae of Inorganic Compounds	P. Fodor-Csányi	1980	* 1984	Commission III.1
5/81	Non-Structural Stoichiometric Names : Addition Compounds	T.D. Coyle	1980	1984	
6/81	Solid State Phases (Revised title)	R. Metselaar	1980	* 1984	Commission II.3
7/81	Neutral Molecular Compounds	E.W. Godly	1980	* 1984	Commission III.1
8/81	Ions, Radicals and Salts	J. Reedijk	1980	* 1984	Commission III.1
9/81	Acids and Derivatives, Including Oxoacids	Y. Jeannin	1980	* 1984	Commission III.1
10/81	Coordination Compounds	D.H. Busch	1980	* 1984	Commission III.1
	New Nomenclature				
11/81	Isopoly and Heteropolyanions	Y. Jeannin	1976	* 1984	
12/81	Inorganic Chain Compounds (Revised title)	E. Fluck	1977	* 1984	Commission III.1

COMMISSION ON NOMENCLATURE OF INORGANIC CHEMISTRY (II.2) (CONTINUED)

	PROJECT TITLE	COORDINATOR	STARTING DATE	COMPLETION OR * REVIEW DATE	OTHER INTERESTS AND COMMENTS
	New Nomenclature				
13/81	Stereochemical Nomenclature	T.E. Sloan	1976	* 1984	Commission III.1
16/81	Red Book Revision, Planning and Methods	G.J. Leigh	1977	1985	
17/83	Nomenclature of Inorganic Ring Compounds	K.Ch. Buschbeck	1983	* 1985	Commission III.1

COMMISSION ON HIGH TEMPERATURE AND SOLID STATE CHEMISTRY (II.3)

	PROJECT TITLE	COORDINATOR	STARTING DATE	COMPLETION OR * REVIEW DATE	OTHER INTERESTS AND COMMENTS
	Project 4/81 completed				
1/81	Thermodynamic and Transport Properties of Alkali Metals	R.W. Ohse	1976	* 1984	Atomic Energy Commissions, IAEA, CODATA
2/81	Bibliography of High Temperature Chemistry and Physics of Materials	M.G. Hocking	1956	NPR Ongoing	Commission I.2
3/81	IUPAC Thermochemical Data Bank	C.B. Alcock	1978	Ongoing	CODATA, Commission I.2
5/81	Melting Points of Refractory Oxides - Part II	M. Rand J. Coutures	1976	* 1984	
6/81	Characterization and Terminology of Carbon	E. Fitzer	1976	* 1984	ASTM, CODATA
7/81	Carbide-Graphite Secondary Temperature Standards (Revised title)	K.L. Komarek	1979	* 1984	
8/81	Properties of Si_3N (Revised title)	R. Sersale G. Petzow	1978	* 1984	

COMMISSION ON HIGH TEMPERATURE AND SOLID STATE CHEMISTRY (II.3) (CONTINUED)

PROJECT	TITLE	COORDINATOR	STARTING DATE	COMPLETION OR * REVIEW DATE	OTHER INTERESTS AND COMMENTS
9/83	High Temperature Mass Spectrometry Ionization Cross Sections	J. Drowart	1982	* 1984	Commission I.5
10/83	Reference Electrodes for Solid State EMF Studies	M.H. Rand	1982	* 1984	Commission I.3
11/83	Nomenclature of Solids	R. Metselaar J. Corish	1981	* 1984	Commission II.2
12/83	Ionic Conductivity of Fluorite-Structured Oxides	A.M. Anthony J.F. Baumard R. Metselaar	1983	* 1984	Commission I.3
13/83	Classification of Si_3N_4 Based Materials	G. Petzow R. Sersale	1983	1985	

ORGANIC CHEMISTRY DIVISION (III)

President: Prof. G. Modena (Istituto di Chimica Organica dell'Università, Via Marzolo 1, I-35100 Padova, Italy)

Vice-President: Prof. J. F. Bunnett

Secretary: Dr. Sukh Dev

COMMISSION ON NOMENCLATURE OF ORGANIC CHEMISTRY (III.1)

	PROJECT TITLE	COORDINATOR	STARTING DATE	COMPLETION OR * REVIEW DATE	OTHER INTERESTS AND COMMENTS
	Projects 11/81 published, 16/83 published under revised title 'Extension of Rules A-1.1 and A-2.5 concerning Numerical Terms used in Organic Chemical Nomenclature' (see Appendix I)				
2/81	Revised Nomenclature for Radicals, Ions and Radical Ions - Provisional	W.H. Powell	1979	1986	Commissions II.2, III.2
3/81	Systematic Nomenclature for Fused Rings - Provisional	A.D. McNaught	1978	1987	
4/81	Nomenclature of Cyclophanes - Provisional	N. Lozac'h	1973	* 1984	
5/81	Acid Nomenclature - Provisional	P.A.S. Smith	1976	1985	Commission II.2
6/81	Revision of 1979 Organic Nomenclature Rules - Section 'R' - Provisional	J.C. Richer	1978	1986	Commissions II.2, III.2, JCBN
7/81	Nomenclature of Organic Chemistry Section 'E' - Stereochemistry - Revised	G.P. Moss	1978	1986	Commissions II.2, III.2, JCBN
8/81	Nomenclature of Organic Chemistry Section 'F' - Natural Products and Related Compounds - Revised	W.H. Powell	1980	1986	JCBN

COMMISSION ON NOMENCLATURE OF ORGANIC CHEMISTRY (III.1) (CONTINUED)

PROJECT	TITLE	COORDINATOR	STARTING DATE	COMPLETION OR * REVIEW DATE	OTHER INTERESTS AND COMMENTS
9/81	Glossary of Class Names for Organic Compounds, Groups and Intermediates definable in Structural Terms (Revised title) - Provisional	P.A.S. Smith	1980	1986	Joint with Commission III.2
10/81	Treatment of Variable Valence in Organic Nomenclature (λ Convention)(Revised title) - Revised	W.H. Powell	1976	1984	Commission II.2
12/83	Systematic Nomenclature for Delocalized Ions and Radicals - Provisional	D. Hellwinkel	1976	1987	Commissions II.2, III.2
13/83	Radial Nomenclature - Provisional	K.H. Hirayama	1980	1985	
14/83	Nomenclature for Cyclic Organic Compounds with Contiguous Formal Double Bonds (δ Convention) (Revised title) - Provisional	W.H. Powell	1980	1984	Commission II.2
15/83	Systematic Names for Mono-nuclear Parent Hydrides in Substitutive Nomenclature - Provisional	W.H. Powell	1981	1985	Commission II.2
17/83	Nodal Numbering - Part I : Acyclic and Cyclic Systems	N. Lozac'h	1982	1984	Commission II.2
18/83	The Technique of Indicated Hydrogen in Organic Nomenclature	A.D. McNaught	1981	1986	
19/83	Nomenclature of Polycyclic Fused and Bridged Fused Ring Systems - Provisional	G.P. Moss	1978	1986	

COMMISSION ON NOMENCLATURE OF ORGANIC CHEMISTRY (III.1) (CONTINUED)

PROJECT	TITLE	COORDINATOR	STARTING DATE	COMPLETION OR * REVIEW DATE	OTHER INTERESTS AND COMMENTS
20/83	Extension of Rules A-1.1 and A-2.5 concerning Numerical Terms used in Organic Chemical Nomenclature - Revised	N. Lozac'h	1983	1985	

COMMISSION ON PHYSICAL ORGANIC CHEMISTRY (III.2)

Projects 3/81 published (see Appendix I), 4.3/81 terminated without publication

PROJECT	TITLE	COORDINATOR	STARTING DATE	COMPLETION OR * REVIEW DATE	OTHER INTERESTS AND COMMENTS
2/81	Nomenclature for Multivalent Transformations	R.A.Y. Jones	1978	* 1984	
	Symbolization of Reaction Mechanisms in Organic Chemistry				
4.1/81	Part A	R.D. Guthrie	1976	1985	
4.2/81	Part B	J.S. Littler	1976	1985	
5/81	Nomenclature for Transformations : Attachment/Detachment (Revised title)	M.P. Doyle	1978	* 1984	
6/81	Nomenclature for Ring-Forming and Ring-Opening Transformations	P. Müller	1978	* 1985	
7/81	Nomenclature for Complex Reactions	J. March	1977	* 1985	
8/83	Nomenclature for Transformations : Coupling/Uncoupling (Revised title)	M.P. Doyle	1978	* 1985	
9/83	Nomenclature for Insertion and Extrusion Transformations	J. Toullec	1981	* 1984	
10/83	Nomenclature for Rearrangement Transformations	R.A.Y. Jones	1981	* 1984	

COMMISSION ON PHYSICAL ORGANIC CHEMISTRY (III.2) (CONTINUED)

PROJECT	TITLE	COORDINATOR	STARTING DATE	COMPLETION OR * REVIEW DATE	OTHER INTERESTS AND COMMENTS
11/83	Stereochemical Terminology	V. Gold	1981	* 1985	Commissions II.2, III.1, IV.1
12/83	Glossary of Terms used in Physical Organic Chemistry - Revised	V. Gold	1982	1986	Commission III.3

COMMISSION ON PHOTOCHEMISTRY (III.3)

PROJECT	TITLE	COORDINATOR	STARTING DATE	COMPLETION OR * REVIEW DATE	OTHER INTERESTS AND COMMENTS
2/81	Glossary of Terms Used in Photochemistry	S.E. Braslavsky	1978	1985	
3/81	Educational Aspects of Photochemistry Concepts	F.C. De Schryver	1979	1985	Divisions I, II, IV, CTC
	Photochemical and Photophysical Standards				
4.1/81	Fluorescence	D.F. Eaton	1981	1985	
5/83	Compilation of Photophysical and Photochemical Data (Revised title)	F. Wilkinson	1981	1987	
	- - - - - - - - Projects approved from August 1983 - - - - - - - -				
6/83	Recommended Standards for Reporting Photophysical and Photochemical Data - Revised	F.C. De Schryver	1982	1984	
7/83	Development of Photochemical Experiments for Undergraduate Laboratories	K. Tokumaru	1983	1987	

COMMISSION ON MEDICINAL CHEMISTRY (III.4) (UNDER REVIEW)

MACROMOLECULAR DIVISION (IV)

President: Prof. C. H. Bamford (Bioengineering and Medical Physics Unit, Duncan Building, Faculty of Medicine, University of Liverpool, Royal Liverpool Hospital, Liverpool L7 8XP, UK)

Vice-President: Prof. T. Saegusa

Secretary: Prof. M. Mandel

COMMISSION ON MACROMOLECULAR NOMENCLATURE (IV.1)

	PROJECT TITLE	COORDINATOR	STARTING DATE	COMPLETION OR * REVIEW DATE	OTHER INTERESTS AND COMMENTS
	Project 7/81 passed to Division III				
3/81	A Classification System for Single-Strand Polymers - Provisional	I.M. Papisov N.M. Bikales	1978	1984	
4/81	Copolymer Nomenclature - Provisional	W. Ring A.D. Jenkins	1977	1984	
	Physical Terms and Definitions				
5.1/81	Solutions - Provisional	P. Kratochvíl U.W. Suter	1978	1985	Division I
5.2/81	Crystalline Polymers - Provisional	G. Allegra	1978	1985	
5.3/81	Melts, Solid State - Provisional	R.E. Wetton A.D. Jenkins	1978	* 1985	
6/81	Subsidiary Terms and Definitions in Polymer Chemistry - Provisional	P. Sigwalt I. Mita N.M. Bikales	1978	* 1984	
8/83	Substitutive Nomenclature - Provisional	R.B. Fox	1979	* 1985	

COMMISSION ON MACROMOLECULAR NOMENCLATURE (IV.1) (CONTINUED)

	PROJECT TITLE	COORDINATOR	STARTING DATE	COMPLETION OR * REVIEW DATE	OTHER INTERESTS AND COMMENTS
9/83	Nomenclature for Interpenetrating Networks, Blends, etc. - Provisional	L.G. Donaruma	1981	* 1984	Consultation with Polymer Nomenclature Committee of ACS
10/83	Nomenclature for Ladder Polymers - Provisional	A.D. Jenkins	1981	* 1985	
	- -	Project approved from August 1983		- - - - - - - - - - - - - - - -	
11/83	Nomenclature Compendium	A.D. Jenkins	1983	1984	

COMMISSION ON POLYMER CHARACTERIZATION AND PROPERTIES (IV.2)

1. WORKING PARTY ON STRUCTURE AND PROPERTIES OF COMMERCIAL POLYMERS

Projects 1.2/81 terminated, 1.4/81 published (see Appendix I)

Elongational Flow of Polymer Melts

	PROJECT TITLE	COORDINATOR	STARTING DATE	COMPLETION OR * REVIEW DATE	OTHER INTERESTS AND COMMENTS
1.7.1/81	The Stability of Elongational Flows of Molten Polyethylene	J.L. White	1980	1985	
1.7.2/81	The Stability of Elongational Flows of Molten Polyethylenterephthalate	J.L. White	1980	1985	
1.7.3/81	Extrusion Instabilities of Polyethylene	J.L. White	1980	1985	
1.8/83	A Study of Runaway Creep in High Impact Polystyrene	J.M. Cann	1979	1984	

1. WORKING PARTY ON STRUCTURE AND PROPERTIES OF COMMERCIAL POLYMERS (CONTINUED)

PROJECT TITLE		COORDINATOR	STARTING DATE	COMPLETION OR * REVIEW DATE	OTHER INTERESTS AND COMMENTS
1.9/83	Application of Fracture Mechanics to the Prediction of the Ductile/Brittle Transition	C.B. Bucknall	1982	* 1984	
1.10/83	Impact Toughness of Polypropylene	C.B. Bucknall	1982	* 1984	
1.11/83	Viscoelastic Properties of Rubber-Modified Polymeric Materials at Elevated Temperatures (Revised title)	T. Masuda	1981	1984	
1.12/83	Relationships between Molecular Characteristics and Physical Properties of Linear Low Density Polyethylene (LLDPE)	T. Okamoto	1982	1985	
1.13/83	Defects in the Molecular Structure of Polyvinylchloride and their Relation to Thermal Stability (Revised title)	A. Michel	1977	* 1984	

2. WORKING PARTY ON MOLECULAR CHARACTERIZATION OF COMMERCIAL POLYMERS

PROJECT TITLE		COORDINATOR	STARTING DATE	COMPLETION OR * REVIEW DATE	OTHER INTERESTS AND COMMENTS
2.3/83	Molecular Characterization of Linear Low Density Polyethylene	P. Starck	1981	1985	
2.4/83	Molecular Characterization of Ethylene-Propylene Block Copolymers	Th.G. Scholte	1979	1984	
2.5/83	Characterization of Polyelectrolytes	M. Mandel	1980	1989	

2. WORKING PARTY ON MOLECULAR CHARACTERIZATION OF COMMERCIAL POLYMERS (CONTINUED)

	PROJECT TITLE	COORDINATOR	STARTING DATE	COMPLETION OR * REVIEW DATE	OTHER INTERESTS AND COMMENTS
	- - - - - - Projects approved from August 1983 - - - - - -				
3/83	Molecular Characterization of Linear Low Density Polyethylene	P. Starck	1983	1985	
4/83	Branching Frequence Distribution of Low Density Polyethylene	P. Starck	1983	1985	

3. WORKING PARTY ON SUPPORTED POLYMER FILMS

	PROJECT TITLE	COORDINATOR	STARTING DATE	COMPLETION OR * REVIEW DATE	OTHER INTERESTS AND COMMENTS
3.1/81	Mechanism and Measurement of Adhesion	H. Kollek	1976	1984	
3.2/81	Analysis of Emulsion Paints	Ch.M. Hansen	1980	* 1984	
3.3/81	Analysis and Stability of Systems Containing Amino Resins	H.J. Luthardt	1980	1985	

4. WORKING PARTY ON THERMAL PROPERTIES OF POLYMERS

	PROJECT TITLE	COORDINATOR	STARTING DATE	COMPLETION OR * REVIEW DATE	OTHER INTERESTS AND COMMENTS
4.1/81	Glass Transition Phenomena of Amorphous Polystyrene by DTA and DSC (Revised title)	G. Hinrichsen	1981	* 1984	

5. WORKING PARTY ON KINETIC PARAMETERS FOR FREE RADICAL POLYMERIZATION

	PROJECT TITLE	COORDINATOR	STARTING DATE	COMPLETION OR * REVIEW DATE	OTHER INTERESTS AND COMMENTS
5.1/81	Kinetic Parameters for Styrene and Methylmethacrylate	G. Eastmond	1979	* 1984	

ANALYTICAL CHEMISTRY DIVISION (V)

President: Prof. F. Pellerin (Faculté de Pharmacie de l'Université de Paris-Sud, Hôpital Corentin-Celton, 37 boulevard Gambetta, F-92130 Issy les Moulineaux, France)

Vice-President: Prof. G.H. Nancollas

Secretary: Prof. G. den Boef

ANALYTICAL CHEMISTRY DIVISION COMMITTEE (V)

PROJECT TITLE		COORDINATOR	STARTING DATE	COMPLETION OR * REVIEW DATE	OTHER INTERESTS AND COMMENTS
1/81	Determination of Trace Metals and Metalloids in Natural Waters (Revised title)	H.W. Nürnberg	1979	* 1984	Division VI
2/81	Multilingual Dictionary of Analytical Terms	E. Pungor	1980	* 1984	
3/81	Compendium of Analytical Nomenclature (2nd Edition) - Revised	H. Freiser	1979	1985	
4/83	Second International Symposium on Harmonization of Collaborative Analytical Studies (Washington, 84.10.25-27)	F. Pellerin	1982	1984	Joint with Divisions VI and VII
	- - - - - - - - - - - - Project approved from August 1983 - - - - - - - - - - - -				
5/83	Trace Metals and Metalloids in Environmental Biological Materials and Food	H.W. Nürnberg	1984	1987	

COMMISSION ON ANALYTICAL REACTIONS AND REAGENTS (V.1)

	PROJECT TITLE	COORDINATOR	STARTING DATE	COMPLETION OR * REVIEW DATE	OTHER INTERESTS AND COMMENTS
	Projects 2/81 and 8/81 published (see Appendix I)				
1/81	Critical Survey of Spectro-photometric and Fluorimetric Methods Used in Water Analysis	G. den Boef	1980	1984	Contribution to Division V Project 1/81
3/81	Colorimetric and Fluorimetric Determination of Amines	M. Pesez J. Bartos	1972	1984	
4/81	Identification and Purity Control of Acid-Base Indicators (This project was re-started in 1981)	G. Ackermann	1981	1984	
9/83	Quantitative Characterization of Spectrophotometric Methods in UV and Visible	G. Ackermann	1981	1984	
10/83	Classification of Interferences	G. Ackermann	1982	1985	
11/83	Metal Buffers in Chemical Analysis	E. Wänninen	1982	1985	
	- - - - - Projects approved from August 1983 - - - - -				
12/83	Characterization of Stationary Phases Used for Gas Chromatography	D. Thorburn Burns	1983	1985	
13/83	Present Status and Perspectives of Organic Analytical Reagents	L. Sommer	1984	1986	
14/83	Terminology, Classification and Presentation of Results of Pattern Recognition Techniques	D.L. Massart	1984	1986	

COMMISSION ON MICROCHEMICAL TECHNIQUES AND TRACE ANALYSIS (V.2)

PROJECT	TITLE	COORDINATOR	STARTING DATE	COMPLETION OR * REVIEW DATE	OTHER INTERESTS AND COMMENTS
	Projects 4/81 published (see Appendix I), 6/81 withdrawn, 12.6/83 abandoned				
1.2/81	General Aspects of Trace Analytical Methods - VI. Acid Pressure Decomposition in Trace Element Analysis	S. Gomiščěk E. Jackwerth	1973	1984	
7.1/81	Separation and Preconcentration of Trace Substances - II. Multielement Preconcentration by Precipitation of the Matrix	E. Jackwerth A. Townshend	1979	1984	
8/81	Standards and Calibration Methods for Surface Analysis - Part I (Revised title)	M. Grasserbauer	1980	* 1985	
	Two reports agreed by Division for publication				
9.1/81	Some General Considerations Concerning Accuracy and Precision in Trace Metal Analysis in Water	B. Griepink	1979	* 1984	Contribution to Division V Project 1/81
9.2/81	Investigations of Methods and Sources of Error in Trace Element Analysis in Surface Waters Using the Determination of Chromium and Mercury as examples	B. Griepink	1979	* 1984	Contribution to Division V Project 1/81
10/83	Methodology of Trace Organic Analysis - I. A Preliminary Survey	A. Lamotte	1981	* 1984	Commission VII.4
11/83	Survey of Decomposition Methods in Elemental Trace Analysis	B. Griepink	1981	1985	
	Critical Evaluation of Analytical Methods for Determination of Trace Elements in Various Matrices				Commission VII.4
12.1/83	Selenium in Biological Materials and Water	A.D. Campbell	1979	* 1984	

COMMISSION ON MICROCHEMICAL TECHNIQUES AND TRACE ANALYSIS (V.2) (CONTINUED)

	PROJECT TITLE	COORDINATOR	STARTING DATE	COMPLETION OR * REVIEW DATE	OTHER INTERESTS AND COMMENTS
	Critical Evaluation of Analytical Methods for Determination of Trace Elements in Various Matrices				Commission VII.4
12.2/ 83	Cobalt in Biological Materials	A. Townshend	1979	* 1984	
12.3/ 83	Molybdenum in Biological Materials	Z. Marczenko	1979	* 1984	
12.4/ 83	Chromium in Biological Materials	J.M. Ottaway	1979	* 1984	
12.5/ 83	Thallium in Various Matrices	G. Tölg B. Griepink	1979	1984	
12.7/ 83	Semi-Conductor Materials (Project divided into 2 Parts, and titles revised)				
	Part A - Bulk Analysis	Yu.A. Zolotov	1981	1984	
	Part B - Distribution Analysis	M. Grasserbauer	1981	1984	
12.8/ 83	Manganese in Biological Materials	J.M. Ottaway	1981	* 1984	
12.9/ 83	Fluorine in Biological Materials	A.D. Campbell	1982	1985	
	- - - - - - - - - - - Projects approved from August 1983 - - - - - - - - - - -				
13/83	Reference Materials for Trace Analysis	B. Griepink	1983	1987	EEC

COMMISSION ON MICROCHEMICAL TECHNIQUES AND TRACE ANALYSIS (V.2) (CONTINUED)

	PROJECT TITLE	COORDINATOR	STARTING DATE	COMPLETION OR * REVIEW DATE	OTHER INTERESTS AND COMMENTS
	Separation and Preconcentration of Trace Substances				
14/83	Part IV - New Type of Flotation of Ion-Association of Multivalent Anionic Complexes with Basic Dyes (Taking Platinum Metals as Examples)	Z. Marczenko	1983	1984	
15/83	Part V - Microscale Preconcentration Techniques for Trace Analysis	A. Mizuike	1983	1987	
16/83	Part VI Aspects of Preconcentration and Determination Methods in Extreme Trace Analysis	Yu.A. Zolotov	1983	1985	
	Critical Evaluation of Analytical Methods for the Determination of Trace Elements in Various Matrices				
17/83	Aluminium in Biological Materials	A. Cedergren	1983	1985	

COMMISSION ON ANALYTICAL NOMENCLATURE (V.3)

	PROJECT TITLE	COORDINATOR	STARTING DATE	COMPLETION OR * REVIEW DATE	OTHER INTERESTS AND COMMENTS
	Project 16/83 abandoned				
1/81	Nomenclature of Kinetic Methods of Analysis	G. Svehla	1977	1984	
2/81	Nomenclature of Sampling	W. Horwitz	1978	1985	Commissions V.4, VII.1, REMCO
3/81	Nomenclature of Information Storage and Retrieval	S.P. Perone	1978	* 1984	Commission VII.1
4/81	Nomenclature of Automated Analysis	G.G. Guilbault	1977	1985	Divisions V, VII, Commission VII.1

COMMISSION ON ANALYTICAL NOMENCLATURE (V.3) (CONTINUED)

PROJECT	TITLE	COORDINATOR	STARTING DATE	COMPLETION OR * REVIEW DATE	OTHER INTERESTS AND COMMENTS
5/81	Presentation of Papers for Publication in Luminescence Spectroscopy (Revised title)	G.G. Guilbault	1978	1984	Commissions I.5, V.4
7/81	Presentation of Results of Chemical Analysis (Revision of Chapter 1 of Compendium)	G. Svehla	1978	1984	Division V, Commissions V.1, V.4, REMCO
8/81	Terminology of Environmental Analysis - Part I Terms Related to Environmental Analytical Methodology (Revised title)	R.W. Frei D. Klockow	1977	1985	
9/81	Nuclear Analysis Nomenclature	R.E. Van Grieken	1977	* 1984	
10/81	Chromatography Nomenclature (Revision of Chapters 13 and 14 of Compendium)	L.S. Ettre C.A.M.G. Cramers	1978	1985	
11/81	Ion-Exchange Nomenclature	M.A. Leonard C.A.M.G. Cramers	1977 (Restarted 1983)	1985	
12/81	Chiroptical Methods Nomenclature	W. Simon	1979	1984	Commission I.5
14/81	Liquid-Liquid Extraction	H.M.N.H. Irving	1977	1984	
15/83	Nomenclature of Bioanalytical Methods - Part I General Nomenclature (Revised into Parts)	G.G. Guilbault	1981	1985	Commission VII.1
17/83	Nomenclature of Chemometrics (Revised title)	S.P. Perone B.R. Kowalski	1981	1985	Commission VII.1, Chemometrics Society (USA), NATO Advanced Study Institute of Chemometrics
	- - - - - - Projects approved from August 1983 - - - - - -				
18/83	Nomenclature in Evaluation of Analytical Methods	W. Horwitz	1984	1987	Commission VII.1, REMCO, ISO, Harmonization Symposium

COMMISSION ON ANALYTICAL NOMENCLATURE (V.3) (CONTINUED)

PROJECT	TITLE	COORDINATOR	STARTING DATE	COMPLETION OR * REVIEW DATE	OTHER INTERESTS AND COMMENTS
19/83	Nomenclature of NMR and EPR Spectroscopy	D. Leyden	1983	1987	Commission I.5
20/83	Nomenclature of Thermometric and Enthalpymetric Methods	D. Klockow	1983	1987	ICTA
21/83	Nomenclature of Electrochemical Separation Techniques	C.A.M.G. Cramers	1984	1987	Commission V.5

COMMISSION ON SPECTROCHEMICAL AND OTHER OPTICAL PROCEDURES FOR ANALYSIS (V.4)

PROJECT	TITLE	COORDINATOR	STARTING DATE	COMPLETION OR * REVIEW DATE	OTHER INTERESTS AND COMMENTS
	Project 1.4.3/81 abandoned				
	Nomenclature, Symbols, Units, and their Usage in Spectrochemical Analysis				
1.3/81	Preparation of Index for Parts I - X (Revised title)	A. Strasheim K. Laqua	1975	1985	
1.4.1/81	Translation into French	J. Robin	1979	* 1985	
1.4.2/81	Translation into German	K. Laqua	1979	* 1985	
1.4.4/81	Translation into Czechoslovakian	E. Plško	1979	* 1985	
1.5/81	Analytical Optical Molecular Absorption Spectroscopy : UV and Visible - Part VII	M. Zander	1979	* 1984	Commissions I.5, V.3, VII.2
1.6/81	Nomenclature System for X-Ray Spectroscopy - Part VIII (Revised title)	C. Sénémaud	1979	1985	IUPAP, IUCr

COMMISSION ON SPECTROCHEMICAL AND OTHER OPTICAL PROCEDURES FOR ANALYSIS (V.4) (CONTINUED)

PROJECT	TITLE	COORDINATOR	STARTING DATE	COMPLETION OR * REVIEW DATE	OTHER INTERESTS AND COMMENTS
1.7/81	Instrumentation for Dispersion and Isolation of Optical Spectra - Part IX	K. Laqua	1979	1984	IUPAP
1.8/81	Preparation of Materials for Analytical Atomic Spectroscopy - Part X	A.M. Ure	1979	* 1984	Commissions V.2, V.3
1.9/83	Radiation Detectors - Part XI	K. Laqua	1981	1985	
2/81	Determination of Trace Elements in Natural Waters by Atomic Spectroscopy Techniques	A. Strasheim	1979	* 1984	Contribution to Division V Project 1/81
	- Projects approved from August 1983 -				
	Nomenclature, Symbols, Units, and their Usage in Spectro-chemical Analysis				
3/83	Electrothermal Atomization and Hydride Generation - Part XII	I. Rubeška B.V. L'vov	1983	1987	Commission V.2
4/83	Atomic Florescence Spectroscopy - Part XIII	J.M. Mermet	1983	1987	IUPAP

COMMISSION ON ELECTROANALYTICAL CHEMISTRY (V.5)

Projects 4/81 Part I, 6/81, 14/81, 18/83 (Revised title) published (see Appendix I)

PROJECT	TITLE	COORDINATOR	STARTING DATE	COMPLETION OR * REVIEW DATE	OTHER INTERESTS AND COMMENTS
4/81	Thermodynamic Constants for Individual Ionic Species - II, III, IV, V	Y. Marcus	1981	* 1985	Commission V.6
5/81	Voltammetry at Glassy Carbon Electrodes	M. Gross	1979	1984	

COMMISSION ON ELECTROANALYTICAL CHEMISTRY (V.5) (CONTINUED)

PROJECT	TITLE	COORDINATOR	STARTING DATE	COMPLETION OR * REVIEW DATE	OTHER INTERESTS AND COMMENTS
7/81	Mediators in Spectroeletrochemistry (Revised title)	R.A. Durst	1979	1985	
8/81	Oxidation-Reduction Potentials in Aqueous Solutions - A Selective and Critical Source Book	J. Jordan	1977	1984	Joint with Commission I.3
9/81	Procedures for Testing pH Responsive Glass Electrodes at 25°, 65° and 85°C and Determination of Alkaline Errors up to 1 mol dm^{-3} Na^+, K^+, Li^+ (Revised title)	A.K. Covington	1977	1984	
12/81	Response Times of Ion-Selective Electrodes and Potentiometric Cells	E. Pungor	1977	1985	
13/81	Electroanalytical Chemistry of Sulfur Compounds in New Coal Conversion Technologies	J. Jordan	1977	1984	
17/83	Standard Systems for Heterogeneous Charge Transfer Rates	K.M. Kadish	1981	* 1985	Commission I.3
19/83	Chemically Modified Electrodes - Provisional	R.A. Durst	1981	* 1984	Commission I.3
20/83	Voltammetry at Interfaces Between Immiscible Solvents - Provisional	M. Senda	1981	1985	
21/83	Criteria for Standardization of pH Measurements in Water-Rich Mixed Solvents (Revised title)	J. Juillard	1981	* 1985	
22/83	Purification of Solvents for Electroanalysis (Revised title)	J.F. Coetzee	1981	* 1985	

COMMISSION ON ELECTROANALYTICAL CHEMISTRY (V.5) (CONTINUED)

PROJECT	TITLE	COORDINATOR	STARTING DATE	COMPLETION OR * REVIEW DATE	OTHER INTERESTS AND COMMENTS
23/83	Acid-Base Dissociation Constants in Dipolar Aprotic Solvents	K. Izutsu	1975	1984	Commissions I.3, V.5
24/83	Standard Potentials of Amalgam Electrodes and Activity Coefficients of Metals in Mercury	T. Mussini	1981	* 1984	Commission I.3
	- - - - - - - - - -	Projects approved from August 1983		- - - - - - - - - -	
25/83	Critical Comparison between Various Carbon Indicator Electrodes in Voltammetry	M. Gross	1983	1987	Commission I.3
26/83	New Definition of Polarography	R. Kalvoda A.A. Vlček	1983	1985	
27/83	Translation into Chinese of Selected Reports of Commission V.5	H. Kao	1983	Ongoing	
28/83	Evaluation of Standard Potentials from e.m.f. Data in Nonaqueous and Mixed Solvents by Extrapolation	Y. Marcus	1983	* 1987	Commission I.3

SUBCOMMITTEE ON ELECTROANALYTICAL METHODS OF ENVIRONMENTAL TRACE ANALYSIS

1.1/83	Determination of Trace Elements in Natural Waters, Drinking Water and Waste Water (Revised title)	H.W. Nürnberg	1981	* 1984	
1.2/83	Electrochemical Methods for Speciation of Trace Elements in Natural Waters (Revised title)	J. Buffle	1981	1984	
1.3/83	Trace Metals in Food and Other Biological Materials (Revised title)	H.W. Nürnberg	1981	* 1984	

SUBCOMMITTEE ON ELECTROANALYTICAL METHODS OF ENVIRONMENTAL TRACE ANALYSIS (CONTINUED)

PROJECT	TITLE	COORDINATOR	STARTING DATE	COMPLETION OR * REVIEW DATE	OTHER INTERESTS AND COMMENTS
1.4/83	Organic Pollutants in Water, Biological Materials and Soils	W.F. Smyth	1981	* 1984	
1.5/83	Trace Elements in Soils	E. Pungor	1981	* 1984	
2/83	Determination of pH in Natural Waters	A.K. Covington	1983	1985	
3/83	Electroanalytical Methods for the Determination of Pesticides	J.G. Osteryoung	1983	1987	Commission VI.5

COMMISSION ON EQUILIBRIUM DATA (V.6)

Project 7/81 merged with 3/81, 8/81 has been terminated

PROJECT	TITLE	COORDINATOR	STARTING DATE	COMPLETION OR * REVIEW DATE	OTHER INTERESTS AND COMMENTS
3/81	Critical Surveys of Stability Constants (Series)	H. Ohtaki	Pre 1976	* 1985	Divisions II, V
6/81	Equilibrium Data for Trace Metal Ligand Interaction in Natural Waters	S. Ahrland	1979	* 1985 Ongoing	Contribution to Division V Project 1/81
9/83	Compilation of Stability Constants	I. Grenthe L.D. Pettit	1981	* 1985	
	- - - - - - - - - - Projects approved from August 1983 - - - - - - - - - -				
10/83	Interpretation of Metal Complex Equilibria with Organic Polyelectrolytes	P. Valenta	1983	1987	
11/83	Rationalization of Ion Exchange Data	E. Högfeldt	1983	1987	
12/83	Ionization Constants of Acids and Bases	D.D. Perrin	1980	* 1985	
13/83	Cyanide Complexes	M.T. Beck	Pre 1976	* 1985	

COMMISSION ON EQUILIBRIUM DATA (V.6) (CONTINUED)

PROJECT	TITLE	COORDINATOR	STARTING DATE	COMPLETION OR * REVIEW DATE	OTHER INTERESTS AND COMMENTS
14/83	Thiocyanate Complexes	G.A. Parke	Pre 1976	* 1985	
15/83	Aminocarboxylate Complexes	A. Georgely	Pre 1976	* 1985	

COMMISSION ON ANALYTICAL RADIOCHEMISTRY AND NUCLEAR MATERIALS (V.7) (UNDER REVIEW)

COMMISSION ON SOLUBILITY DATA (V.8)

Project 1.2/81 published (see Appendix I)

PROJECT	TITLE	COORDINATOR	STARTING DATE	COMPLETION OR * REVIEW DATE	OTHER INTERESTS AND COMMENTS
1.4/81	Hydrogen Halides in Nonaqueous Solvents	H.L. Clever W. Gerrard	1976	NPR	
2/81	Silver, Copper and Gold Halates	H. Miyamoto E.M. Woolley	1979	NPR	

SUBCOMMITTEE ON GAS SOLUBILITIES

PROJECT	TITLE	COORDINATOR	STARTING DATE	COMPLETION OR * REVIEW DATE	OTHER INTERESTS AND COMMENTS
1/83	Solubility of Gases in Molten Salts	R.P.T. Tomkins		1984	
2/83	Ammonia, Amines, Phosphide, Arsine, Stribine (Gas Solubilities in all Solvents)	P.G.T. Fogg C.L. Young		1984	
3/83	Propane, Butane, 2-Methylpropane (Gas Solubilities in all Solvents)	W. Hayduk		1984	

SUBCOMMITTEE ON LIQUID SOLUBILITIES

PROJECT	TITLE	COORDINATOR	STARTING DATE	COMPLETION OR * REVIEW DATE	OTHER INTERESTS AND COMMENTS
1/83	Hydrocarbons with Water and Seawater	D.G. Shaw		1984	
2/83	Halobenzenes and Halophenols with Water	A.L. Horvath F.W. Getzen		1984	

SUBCOMMITTEE ON LIQUID SOLUBILITIES (CONTINUED)

PROJECT	TITLE	COORDINATOR	STARTING DATE	COMPLETION OR * REVIEW DATE	OTHER INTERESTS AND COMMENTS
3/83	Alcohols with Water	A.F.M. Barton		1984	

SUBCOMMITTEE ON SOLID SOLUBILITIES

PROJECT	TITLE	COORDINATOR	STARTING DATE	COMPLETION OR * REVIEW DATE	OTHER INTERESTS AND COMMENTS
1/83	Solubilities of β-Lactams	E. Tomlinson A. Regosz	1978	1984	
2/83	Solubilities of Alkali Metal Halates	H. Miyamoto M. Salomon	1979	1984	
3/83	Solubilities of Metals in Mercury	Z. Galus C. Guminski C. Hirayama	1978	1984	
4/83	Solubility of Lanthanide Halides in Nonaqueous Solvents	T. Mioduski M. Salomon	1978	1984	
5/83	Beryllium, Strontium, Barium, and Radium Sulfates Solubilities	J.W. Lorimer C. Kalidas	1978	1984	

APPLIED CHEMISTRY DIVISION (VI)

President: Dr. H. Frehse (Bayer AG, Sparte Pflanzenschutz, Anwendungstechnik CE, Metabolismus und Rückstände, Gebäude W 7, D-5090 Leverkusen-Bayerwerk, Federal Republic of Germany)

Vice-President: Dr. P. C. Kearney

Secretary: Dr. L. E. Coles

APPLIED CHEMISTRY DIVISION COMMITTEE (VI)

PROJECT TITLE		COORDINATOR	STARTING DATE	COMPLETION OR * REVIEW DATE	OTHER INTERESTS AND COMMENTS
1/83	Second International Symposium on the Harmonization of Collaborative Analytical Studies (Washington, 84.10.25-27)	H. Egan	1982	1984	Joint with Divisions V, VII

COMMISSION ON FOOD CHEMISTRY (VI.1)

Projects 3/81 and 12/81 abandoned, 13/81 published JAOAC, 14/81 Results published in Symposium Proceedings (PAC 1980, 52 (1) 175-181), 18/81 Article published in CI October 1983.

2/81	N-Heterocyclic Carcinogens in Food	G. Grimmer	1977	1984	
5/81	Method of Analysis of Total N-Nitroso Compounds (Revised title)	M. Castegnaro	1980	* 1984	Commission VI.3
11/81	Check Sample Survey for Mycotoxins - III	M. Castegnaro M. Freisen	1979	* 1985	
16/81	Determination of Ochratoxin Residues in Meat	P. Krogh	1979	* 1984	
19/81	Determination of Selenium in Food	L.E. Coles P.L. Schuller	1979	* 1984	

COMMISSION ON FOOD CHEMISTRY (VI.1) (CONTINUED)

PROJECT	TITLE	COORDINATOR	STARTING DATE	COMPLETION OR * REVIEW DATE	OTHER INTERESTS AND COMMENTS
21/81	Determination of Fluoride in Food (at a level of 1-5 ng/g)	T. Kato	1979	* 1984	
25/81	Determination of Mercury in Food	K. Kojima T. Kato	1980	* 1984	
26/83	Multitrichothecene Analysis in Wheat Grains (Revised title)	Y. Ueno	1982	* 1984	Commission VI.5
27/83	Rapid Pre-Selection Method for Benzo(a)Pyrene and Glass-Capillary-Gas-Chromatographic Profile Analysis of Polycyclic Aromatic Compounds in Smoked Foods (Revised title)	G. Grimmer	1982	* 1984	Commission VI.3

COMMISSION ON BIOTECHNOLOGY (VI.2)

PROJECT	TITLE	COORDINATOR	STARTING DATE	COMPLETION OR * REVIEW DATE	OTHER INTERESTS AND COMMENTS
	Bioconversion of Renewable Resources - Proposed Guidelines for Methodologies				
4/81	Measurement of Cellulase Activity	T.K. Ghose	1979	1984	
	Guidelines for Testing Quality and Safety of Live Microbial Products for Agricultural Use				
5/83	Fungal Products	R.C. Righelato	1981	1984	Liaison with Commission VI.5
	- - - - - - Projects approved from August 1983 - - - - - -				
6/83	Recommendations for Micro-biological Quality Specifications for Bacterial Insecticides	R.C. Righelato	1983	* 1984	WHO, IOBC, EEC

COMMISSION ON BIOTECHNOLOGY (VI.2) (CONTINUED)

PROJECT	TITLE	COORDINATOR	STARTING DATE	COMPLETION OR * REVIEW DATE	OTHER INTERESTS AND COMMENTS
7/83	Microbial Biomass for Animal Feed Derived from Agricultural Residues Using Mixed Cultures	M. Ringpfeil	1983	1985	
8/83	Measurement of Hemicellulose	T.K. Ghose	1984	1985	

COMMISSION ON OILS, FATS, AND DERIVATIVES (VI.3)

Projects 2/81 partially completed and absorbed into 28/83, 9/81 abandoned

PROJECT	TITLE	COORDINATOR	STARTING DATE	COMPLETION OR * REVIEW DATE	OTHER INTERESTS AND COMMENTS
6/81	Identification and Determination of Emulsifiers (Derived from Fatty Acids) in Oils, Fats, Fat Products, and Cosmetics - Part I	H. Brüschweiler	1979	1986	
7/81	Publication of 6th Edition of Methods - Second Supplement	A. Hautfenne	1979	* 1984	
15/81	Determination of Polycyclic Aromatic Hydrocarbons	G. Ostermann	1979	* 1985	Commission VI.1
16/81	Determination of Erythrodiol in Refined Grape Seed and Olive Oils	J. Gracian-Tous	1979	1984	
17/81	Total and Quantitative Determination of Sterols in Vegetable Oils and Animal Fats	M. Naudet	1979	1984	
18/81	Determination of Mineral Oil Residues in Oils and Fats	P.W. Hendrikse	1979	1986	
19/81	Extension of Available Methods of Analysis for Fats and Oils to Phospholipids (Commercial Lecithins)	R. Ohlson	1980	1986	
20/81	Determination of Total Phospholipids and Insoluble Impurities in Commercial Lecithins	Ö. Levin	1980	1985	

COMMISSION ON OILS, FATS, AND DERIVATIVES (VI.3) (CONTINUED)

PROJECT	TITLE	COORDINATOR	STARTING DATE	COMPLETION OR * REVIEW DATE	OTHER INTERESTS AND COMMENTS
21/81	Qualitative and Quantitative Determination of Phospholipids and Determination of Phosphorus Content in Commercial Lecithins	J.L. Beare-Rogers	1980	1986	
22/81	Determination of Thiobarbituric Acid Value in Deodorized Vegetable Oils	J. Pokorný	1981	1985	
23/81	Determination of Triglycerides Composition by GLC (Animal Fats and Hydrogenated Oils)	W.D. Pocklington	1981	1984	
24/81	Determination of Solvent Residues in Oils	J.P. Wolff	1981	1985	
25/81	Determination of Solvent Residues in Oilseed Cakes by Headspace GLC	J.P. Wolff	1981	1985	
26/83	Determination of Linolenic and Other (n-3) Fatty Acids in Dietary Fats	J.L. Beare-Rogers G. Hølmer	1983	1987	
	- - - - - - - - - - - -	Projects approved	from August 1983	- - - - - - - - - - - -	
27/83	Determination of Butyric Acid	W.D. Pocklington	1983	1985	IDF, OICC
28/83	Determination of Tocopherols by HPLC	W.D. Pocklington	1983	1986	
29/83	Determination of Metals in Oils and Fats by Atomic Absorption Spectrometry	P.W. Hendrikse	1984	1987	

COMMISSION ON ATMOSPHERIC ENVIRONMENT (VI.4) (UNDER REVIEW)

COMMISSION ON PESTICIDE CHEMISTRY (VI.5)

PROJECT	TITLE	COORDINATOR	STARTING DATE	COMPLETION OR * REVIEW DATE	OTHER INTERESTS AND COMMENTS
7/81	Definition and Description of Non-Extractable Residues in Plants and Soil	T. Roberts	1979	1984	JMPR, CAC
8.1/81	Critical Evaluation of Model Ecosystems	W. Klein	1979	1984	JMPR, CAC
8.2/81	Recommended Approaches to the Evaluation of the Environmental Behaviour of Pesticides (Revised title)	H. Esser	1979	* 1984	JMPR, CAC
15/83	Application of Multi-Residue Methods for Analysis of Samples of Unknown Origin	A. Ambrus	1982	1985	FAO, JMPR
16/83	The Potential Contamination of Groundwater by Pesticides (Revised title)	N. Aharonson	1982	1985	SCOPE
17/83	Recommended Approaches to the Validation of Animal Metabolism in Relation to Man (Revised title)	J. Miyamoto	1982	* 1984	FAO, JMPR
18/83	Interpretation and Optimum Use of Residues Data (Revised title)	J.A.R. Bates S. Gorbach	1982	1985	FAO, JMPR
	- - - - - - - - Projects approved from August 1983 - - - - - - - -				
19/83	Recommended Approaches to the Measurement of Exposure of Workers to Pesticides	R. Hemingway	1983	1985	WHO, ILO

COMMISSION ON PESTICIDE CHEMISTRY (VI.5) (CONTINUED)

PROJECT TITLE		COORDINATOR	STARTING DATE	COMPLETION OR * REVIEW DATE	OTHER INTERESTS AND COMMENTS
20/83	Microbial Adaptation to Pesticides	P.C. Kearney	1983	* 1985	FAO, Commission VI.2
21/83	Confirmatory Techniques for Pesticide Residues	R. Greenhalgh	1983	* 1984	FAO, WHO, Codex Committee on Pesticide Residues

COMMISSION ON WATER QUALITY (VI.6) (UNDER REVIEW)

CLINICAL CHEMISTRY DIVISION (VII)

President: Prof. F. W. Sunderman, Jr. (Departments of Laboratory Medicine and Pharmacology, University of Connecticut Medical School, 263 Farmington Avenue, Farmington, Connecticut 06032, USA)

Vice-President: Dr. S. S. Brown

Secretary: Prof. N. Montalbetti

CLINICAL CHEMISTRY DIVISION COMMITTEE (VII)

PROJECT		COORDINATOR	STARTING DATE	COMPLETION OR * REVIEW DATE	OTHER INTERESTS AND COMMENTS
1/83	Second International Symposium on the Harmonization of Collaborative Analytical Studies (Washington, 84.10.25-27)	F.W. Sunderman, Jr.	1982	1984	Joint with Divisions V, VI

COMMISSION ON AUTOMATION AND CLINICAL CHEMICAL TECHNIQUES (VII.1)

PROJECT		COORDINATOR	STARTING DATE	COMPLETION OR * REVIEW DATE	OTHER INTERESTS AND COMMENTS
1/81	Nomenclature for Automated Analysis in Clinical Chemistry	M. Hjelm	1977	1984	Commission V.3
2/81	Electronic Data Processing and Microprocessors in Clinical Chemistry	P.A. Bonini	1981	1986	Commission V.3
3/83	Bioanalytical Nomenclature	R. Haeckel C.A. Burtis	1982	1986	Commission V.3

COMMISSION ON QUANTITIES AND UNITS (VII.2)

Project 5/81 being finalized by Commission V.4

PROJECT		COORDINATOR	STARTING DATE	COMPLETION OR * REVIEW DATE	OTHER INTERESTS AND COMMENTS
2/81	Nomenclature of Derived Quantities - Provisional	H.P. Lehmann	1976	1984	IFCC

COMMISSION ON QUANTITIES AND UNITS (VII.2) (CONTINUED)

PROJECT	TITLE	COORDINATOR	STARTING DATE	COMPLETION OR * REVIEW DATE	OTHER INTERESTS AND COMMENTS
3/81	Definitions, Symbols, Quantities and Units in Optical Spectroscopy - Provisional	H.P. Lehmann	1980	1984	Commissions I.5, V.5, VI.3
6/81	Nebulizer and Flame Properties in Flame Emission and Absorption Spectroscopy - Provisional	H.P. Lehmann	1974	1984	IFCC
7/81	Quantities Related to Activities and Activity Coefficients - Revised	O. Siggaard-Andersen	1974	1984	IFCC Expert Panel on pH and Blood Gases
10/81	A Special Unit for the Quantity : Number (of Entities) (Revised title)	B.F. Visser	1979	1985	IFCC, ICSH
11/81	Quantities and Units for Metabolic Processes as a Function of Time (Revised title) - Provisional	G. Férard H.P. Lehmann	1979	1986	IFCC
	- - - - - - - - - - Project approved from August 1983 - - - - - - - - - -				
12/83	Quantities and Units in Clinical Chemistry : Classification, Quantities and Units for 'Binding Assays' in the Clinical Laboratory	D.R. Bangham	1983	1986	WHO, IFCC

COMMISSION ON TEACHING (VII.3)

Projects 1/81 Published (see Appendix I), 2/81, 3/81, 8/81 completed without publication, 4/81, 10/81 terminated, 5/81 replaced by 12/83

PROJECT	TITLE	COORDINATOR	STARTING DATE	COMPLETION OR * REVIEW DATE	OTHER INTERESTS AND COMMENTS
6/81	A Basic Education and Training Framework for Medical Laboratory Technicians in Clinical Chemistry (Revised title)	H.G.J. Worth	1978	1984	

COMMISSION ON TEACHING (VII.3) (CONTINUED)

PROJECT	TITLE	COORDINATOR	STARTING DATE	COMPLETION OR * REVIEW DATE	OTHER INTERESTS AND COMMENTS
7/81	Certification, Licensure and Accreditation Programmes for Clinical Chemistry	M.K. Schwartz	1978	1984	
9/81	Preparation of Reference Catalog of Audiovisual and Bibliographic Material for Teaching of Clinical Chemistry	M.K. Schwartz O. Zinder	1979	Ongoing	
	- - - - - Projects approved from August 1983 - - - - -				
11/83	Teaching Guidelines for Clinical Chemists for Effective Communication of Laboratory Data	C.G. Fraser	1983	1985	IFCC, National Societies of Clinical Chemistry
12/83	Guidelines for Teaching Clinical Chemistry to Medical Students	C.G. Fraser O. Zinder	1983	1987	IUB, IFCC, National Societies of Clinical Chemistry
13/83	Drug Effects in Clinical Chemistry; Educational Aspects	H.G.J. Worth	1983	1986	IFCC Expert Panel on Drug Effects

COMMISSION ON TOXICOLOGY (VII.4)

SUBCOMMITTEE ON ENVIRONMENTAL AND OCCUPATIONAL TOXICOLOGY OF NICKEL

PROJECT	TITLE	COORDINATOR	STARTING DATE	COMPLETION OR * REVIEW DATE	OTHER INTERESTS AND COMMENTS
2/81	Reference Method for Analysis of Nickel in Whole Blood and Tissues by Electrothermal AAS	F.W. Sunderman, Jr.	1980	1985	Commission V.2
3/81	Collaborative Harmonization of Nickel Analysis in Body Fluids by Interlaboratory Surveys	S.S. Brown	1977	1985	Commission V.2
4/81	Reference Biological Materials for Nickel Analysis	S.S. Brown	1980	1985	Commission V.2

SUBCOMMITTEE ON ENVIRONMENTAL AND OCCUPATIONAL TOXICOLOGY OF NICKEL (CONTINUED)

PROJECT TITLE		COORDINATOR	STARTING DATE	COMPLETION OR * REVIEW DATE	OTHER INTERESTS AND COMMENTS
5/81	Recommendations for Collection, Storage and Transport of Body Fluids for Nickel Analysis	E. Nieboer	1980	1984	Commission V.2

SUBCOMMITTEE ON ENVIRONMENTAL AND OCCUPATIONAL TOXICOLOGY OF CADMIUM

PROJECT TITLE		COORDINATOR	STARTING DATE	COMPLETION OR * REVIEW DATE	OTHER INTERESTS AND COMMENTS
1/81	Interlaboratory Survey of Blood Cadmium Analysis	D.B. Tonks M. Stoeppler	1978	1984	Commission V.2
2/81	Reliable Methods for Cadmium Analysis in Body Fluids (Revised title)	M. Stoeppler	1980	1986	Commission V.2
3/81	Beta-2 Microglobulin as an Index of Cd-Nephrotoxicity	R.F.M. Herber	1981	1984	Commission V.2

SUBCOMMITTEE ON TOXICOLOGY OF ALUMINIUM

PROJECT TITLE		COORDINATOR	STARTING DATE	COMPLETION OR * REVIEW DATE	OTHER INTERESTS AND COMMENTS
1/83	Reference Method and Materials for Analysis of Aluminium in Biological Materials	J. Savory	1981	1985	Commission V.2

SUBCOMMITTEE ON TOXICOLOGY OF SELENIUM

PROJECT TITLE		COORDINATOR	STARTING DATE	COMPLETION OR * REVIEW DATE	OTHER INTERESTS AND COMMENTS
1/83	Reference Materials and Methods for Analysis of Selenium in Biological Materials	Y. Thomassen	1982	1985	Commission V.2
	- -	Project approved from August 1983		- -	
2/83	The Evaluation of the Hydride Generation Technique in the Determination of Selenium in Human Serum by Atomic Absorption Spectroscopy	B. Welz G. Alfthan	1983	1984	Commission V.2

APPENDIX I — REPORTS PUBLISHED 1983

This listing covers only those activities of Commissions and special Committees which have been completed by publication of a report, normally in the official journal of IUPAC, PAC, or in bookform on behalf of the Union by Pergamon Press.

A listing for the period 1974-80 was included as an appendix to the May 1981 issue of IUPAC Current Programmes and for 1981-2 in the May 1983 issue.

In case of doubt about where to obtain a particular report, please contact the IUPAC Secretariat for further details.

CHEMRAWN COMMITTEE

CHEMRAWN II - Chemistry and World Food Supplies: The New Frontiers - Manila, Phillipines, 6-10 December 1982:

Perspectives and Recommendations: American Chemical Society (International Activities Office), 1983, xii + 28

Invited Papers: Pergamon Press, 1983, xvi + 664

COMMITTEE ON TEACHING OF CHEMISTRY (CTC)

A Laboratory Manual for Low-Budget Teaching of Chemistry - Vol. I: Low-cost Instrumentation Based on Operational Amplifiers for Potentiometry and Conductometry with Illustrative Experiments: UNESCO, 1983, xi + 82

International Newsletter on Chemical Education:

No. 19, July 1983

IUB-IUPAC JOINT COMMISSION ON BIOCHEMICAL NOMENCLATURE (JCBN)

Nomenclature of Retinoids - Provisional: PAC 1983, 55 (4), 721-726

Symbols for Specifying the Conformation of Polysaccharide Chains - Provisional: PAC 1983, 55 (8), 1269-1272

Abbreviations and Symbols for the Description of Conformations of Polynucleotide Chains - Provisional: PAC 1983, 55 (8), 1273-1280

PHYSICAL CHEMISTRY DIVISION (I)

COMMISSION ON ELECTROCHEMISTRY (I.3)

Interphases in Systems of Conducting Phases - Provisional: PAC 1983, 55 (8), 1251-1268

Definition of pH Scales, Standard Reference Values, Measurement of pH and Related Terminology - Provisional: PAC 1983, 55 (9), 1467-1476

COMMISSION ON PHYSICOCHEMICAL MEASUREMENTS AND STANDARDS (I.4)

An Annotated Bibliography on Accuracy in Measurement: PAC 1983, 55 (6), 907-930

COMMISSION ON COLLOID AND SURFACE CHEMISTRY INCLUDING CATALYSIS (I.6)

Definitions, Terminology and Symbols in Colloid and Surface Chemistry - Part 1.14: Light Scattering - Provisional: PAC 1983, 55 (6), 931-942

INORGANIC CHEMISTRY DIVISION (II)

COMMISSION ON ATOMIC WEIGHTS AND ISOTOPIC ABUNDANCES (II.1)

Atomic Weights of the Elements 1981: PAC 1983, 55 (7), 1101-1118

Isotopic Compositions of the Elements 1981: PAC 1983, 55 (7), 1119-1136

COMMISSION ON HIGH TEMPERATURE AND SOLID STATE CHEMISTRY (II.3)

Bibliography on the High Temperature Chemistry and Physics of Materials:

1983, Vol. 27, Nos. 1-4

ORGANIC CHEMISTRY DIVISION (III)

COMMISSION ON NOMENCLATURE OF ORGANIC CHEMISTRY (III.1)

Revision of the Extended Hantzsch-Widman System of Nomenclature for Heteromonocycles: PAC 1983, 55 (2), 409-416

Extension of Rules A-1.1 and A-2.5 Concerning Numerical Terms Used in Organic Chemical Nomenclature - Provisional: PAC 1983, 55 (9), 1463-1466

COMMISSION ON PHYSICAL ORGANIC CHEMISTRY (III.2)

Glossary of Terms Used in Physical Organic Chemistry: PAC 1983, 55 (8), 1281-1371

MACROMOLECULAR DIVISION (IV)

COMMISSION ON POLYMER CHARACTERIZATION AND PROPERTIES (IV.2)

Influence of Acrylic Processing Aids on the Rheology and Structure of Polyvinyl Chloride: PAC 1983, 55 (1), 177-190

A Collaborative Study on the Relation between Film Blowing Performance and Rheological Properties of Two Low-Density and Two High-Density Polyethylene Samples: PAC 1983, 55 (6), 943-976

ANALYTICAL CHEMISTRY DIVISION (V)

COMMISSION ON ANALYTICAL REACTIONS AND REAGENTS (V.1)

Recommendations for the Usage of Selective, Selectivity and Related Terms in Analytical Chemistry: PAC 1983, 55 (3), 553-556

Compleximetric Indicators - Characteristics and Applications: PAC 1983, 55 (7), 1137-1230

COMMISSION ON MICROCHEMICAL TECHNIQUES AND TRACE ANALYSIS (V.2)

Nomenclature, Symbols and Units Recommended for "in situ" Microanalysis - Provisional: PAC 1983, 55 (12), 2023-2027

COMMISSION ON ELECTROANALYTICAL CHEMISTRY (V.5)

Thermodynamic Functions of Transfer of Single Ions from Water to Nonaqueous and Mixed Solvents: Part 1 - Gibbs Free Energies of Transfer to Nonaqueous Solvents: PAC 1983, 55 (6), 977-1022

Half-wave Potentials of Electroactive Substances in Hexamethylphosphoric Triamide: PAC 1983, 55 (8), 1373-1380

Definition of pH Scales, Standard Reference Values, Measurement of pH and Related Terminology - Provisional: PAC 1983, 55 (9), 1467-1476

Applications of Ion-selective Electrodes in Non-aqueous and Mixed Solvents: PAC 1983, 55 (12), 2029-2065

COMMISSION ON EQUILIBRIUM DATA (V.6)

Critical Survey of Stability Constants of Complexes of Indium: PAC 1983, 55 (9), 1477-1528

COMMISSION ON ANALYTICAL RADIOCHEMISTRY AND NUCLEAR MATERIALS (V.7)

Methods for the Determination of Fissile and Fertile Elements - Basic Data of Interest for the Spectrophotometric Determination of Thorium: PAC 1983, 55 (7), 1231-1238

COMMISSION ON SOLUBILITY DATA (V.8)

Solubility Data Series:

- Vol. 12 - Sulfur Dioxide, Chlorine, Fluorine and Chlorine Oxides: Pergamon Press, 1983, xvii + 477
- Vol. 13 - Scandium, Yttrium, Lanthanum and Lanthanide Nitrates: Pergamon Press, 1983, xxiv + 490
- Vol. 14 - Alkaline Earth Metal Halates: Pergamon Press, 1983, xx + 332

APPLIED CHEMISTRY DIVISION (VI)

COMMISSION ON FOOD CHEMISTRY (VI.1)

Recommended Method for the Gas Chromatographic Profile Analysis of Basic N-Containing Aromatic Components (Azaarnes) in High Protein Foods: PAC 1983, 55 (12), 2067-2071

COMMISSION ON OILS, FATS AND DERIVATIVES (VI.3)

Results of a Collaborative Study on Determination of Polar Compounds in Frying Fats: PAC 1983, 55 (8), 1381-1385

COMMISSION ON ATMOSPHERIC ENVIRONMENT (VI.4)

Proposal of Methods for the Determination of Vinyl Chloride and Benzene in Workplace Atmospheres: PAC 1983, 55 (6), 1023-1034

Generating Test Gases for Calibration of Methods Used in Environmental Analysis: PAC 1983, 55 (7), 1239-1250

COMMISSION ON WATER QUALITY (VI.6)

Recommended Notation for Use in the Description of Biological Wastewater Treatment Processes: PAC 1983, 55 (6), 1035-1040

CLINICAL CHEMISTRY DIVISION (VII)

COMMISSION ON AUTOMATION AND CLINICAL CHEMICAL TECHNIQUES (VII.1)

Characteristics and Attributes of Instruments Intended for Automated Analysis in Clinical Chemistry: PAC 1983, 55 (6), 1041-1048

COMMISSION ON TEACHING OF CLINICAL CHEMISTRY (VII.3)

A Scheme for a Two Year Postgraduate Course in Clinical Chemistry: PAC 1983, 55 (3), 557-564

APPENDIX II — SPONSORED SYMPOSIA

(STARTING 1 SEPTEMBER 1983)

(as at 17th April 1984)

This listing starts from 1st September 1983 (i.e. after the conclusion of the 32nd IUPAC General Assembly), and it indicates which Division(s) supported the granting of sponsorship by IUPAC. For further details, please consult CI.

IUPAC CONGRESSES

30th IUPAC Congress
Manchester, UK: 9th-13th September 1985

31st IUPAC Congress
Sofia, Bulgaria: 1987

CHEMRAWN COMMITTEE

CHEMRAWN III: World Conference on Resource Material Conversion: (Bio-)Chemical Process Bridges to meet Future Needs
The Hague, Netherlands: 25th-29th June 1984

CHEMRAWN IV: Chemical Resources of the Global Oceans
Woods Hole/Massachusetts, USA: 22nd-28th September 1985

COMMITTEE ON TEACHING OF CHEMISTRY

8th International Conference on Chemical Education: Widening the Scope of Chemistry
Tokyo, Japan: 18th-23rd August 1985

PHYSICAL CHEMISTRY DIVISION (I)

International Conference on Fourier Transform Spectroscopy
Durham, UK: 5th-9th September 1983

7th International Conference on Crystal Growth
Stuttgart, FRG: 12th-16th September 1983

International Conference on Applications of Mössbauer Effect
Alma Ata, USSR: 26th September-1st October 1983

International Conference on Physics and Chemistry of Low-dimensional Synthetic Metals
Abano Terme, Italy: 17th-22nd June 1984

8th International Congress on Catalysis
Berlin, Federal Republic of Germany: 2nd-6th July 1984

IUPAC Conference on Chemical Thermodynamics
Hamilton, Canada: 13th-17th August 1984

International Conference on Defects in Insulating Crystals
Salt Lake City, USA: 20th-24th August 1984

1st International Symposium on Solubility Phenomena
London, Canada: 21st-23rd August 1984

International Symposium on Physical Chemistry of Colloids and Macromolecules
Uppsala, Sweden: 22nd-24th August 1984

3rd International Conference on Solid Films and Surfaces
Sydney, Australia: 27th-31st August 1984

9th International Conference on Raman Spectroscopy
Tokyo, Japan: 27th August-1st September 1984

10th International Symposium on Reactivity of Solids
Dijon, France: 27th August-1st September 1984

10th International Symposium on Mass Spectrometry
Swansea, UK: 19th-23rd September 1985

INORGANIC CHEMISTRY DIVISION (II)

International Conference on Phosphorus Chemistry,
Nice, France: 5th-9th September 1983

11th International Conference on Organometallic Chemistry
Pine Mountain/Georgia, USA: 10th-14th October 1983

4th International Conference on High Temperature and Energy-Related Materials
Santa Fe/New Mexico, USA: 2nd-6th April 1984

23rd International Conference on Coordination Chemistry
Boulder/Colorado, USA: 29th July-3rd August 1984

9th International Conference on Non-aqueous Solutions
Pittsburgh/Pennsylvania, USA: 13th-17th August 1984

International Conference on Defects in Insulating Crystals
Salt Lake City, USA: 20th-24th August 1984

1st International Symposium on Solubility Phenomena
London, Canada: 21st-23rd August 1984

7th International Symposium on Organosilicon Chemistry
Kyoto, Japan: 9th-14th September 1984

7th International Symposium on Solute-Solute-Solvent Interactions
Reading, UK: 15th-19th July 1985

ORGANIC CHEMISTRY DIVISION (II)

12th International Carbohydrate Symposium
Utrecht, Netherlands: 1st-7th July 1984

14th International Symposium on Chemistry of Natural Products
Poznań, Poland: 9th-14th July 1984

10th IUPAC Symposium on Photochemistry
Interlaken, Switzerland: 22nd-27th July 1984

7th IUPAC Conference on Physical Organic Chemistry
Auckland, New Zealand: 20th-24th August 1984

5th International Conference on Organic Synthesis
Freiburg i. Br., FRG: 27th-30th August 1984

7th International Symposium on Carotenoids
Munich, FRG: 27th-31st August 1984

8th International Symposium on Medicinal Chemistry
Uppsala, Sweden: 27th-31st August 1984

7th International Symposium on Organosilicon Chemistry
Kyoto, Japan: 9th-14th September 1984

3rd IUPAC Symposium on Organometallic Chemistry directed toward Organic Synthesis
Kyoto, Japan: 12th-15th July 1985

5th International Symposium on Novel Aromatic Compounds
St. Andrews, UK: 15th-19th July 1985

15th International Symposium on Chemistry of Natural Products
The Hague, Netherlands: 17th-23rd August 1986

8th IUPAC Conference on Physical Organic Chemistry
Tokyo, Japan: 24th-29th August 1986

MACROMOLECULAR DIVISION (IV)

29th IUPAC International Symposium on Macromolecules
Bucharest, Romania: 5th-9th September 1983

International Conference on Physics and Chemistry of Low-dimensional Synthetic Metals
Abano Terme, Italy: 17th-22nd June 1984

6th Bratislava International Conference on Modified Polymers
Bratislava, Czechoslovakia: 2nd-5th July 1984

26th Prague Microsymposium on Macromolecules: Polymers in Medicine and Biology
Prague, Czechoslovakia: 9th-12th July 1984

27th Prague Microsymposium on Macromolecules: Physical Optics of Dynamic Phenomena and Processes in Macromolecular Systems
Prague, Czechoslovakia: 16th-19th July 1984

International Symposium on Physical Chemistry of Colloids and Macromolecules
Uppsala, Sweden: 22nd-24th August 1984

Polymer Chemistry Meeting
Leuven, Belgium: 26th-28th September 1984

International Symposium on Characterization and Analysis of Polymers
Melbourne, Australia: 11th-14th February 1985

International Symposium on Non Crystalline Order in Polymers
Naples, Italy: 27th-30th May 1985

28th Microsymposium on Macromolecules: Polymer Composites
Prague, Czechoslovakia: 8th-11th July 1985

30th IUPAC Symposium on Macromolecules
The Hague, Netherlands: 18th-23rd August 1985

ANALYTICAL CHEMISTRY DIVISION (V)

1st International Symposium on Solubility Phenomena
London, Canada: 21st-23rd August 1984

International Symposium on Quantitative Luminescence Spectrometry in Biomedical Sciences
Ghent, Belgium: 3rd-6th September 1984

2nd International Symposium on Harmonization of Collaborative Analytical Studies
Washington, DC, USA: 25th-27th October 1984

2nd International Symposium on Analytical Chemistry in Exploration, Mining and Processing of Materials
Pretoria, Republic of South Africa: 15th-19th April 1985

8th International Conference on Thermal Analysis
Bratislava, Czechoslovakia: 19th-23rd August 1985

APPLIED CHEMISTRY DIVISION (VI)

7th International Biotechnology Symposium
New Delhi, India: 19th-25th February 1984

2nd International Symposium on Harmonization of Collaborative Analytical Studies
Washington, DC, USA: 25th-27th October 1984

6th International Symposium on Mycotoxins and Phycotoxins
Pretoria, Republic of South Africa: 22nd-25th July 1985

6th International Congress on Pesticide Chemistry
Ottawa, Canada: 10th-17th August 1986

CLINICAL CHEMISTRY DIVISION (VII)

International Symposium on Quantitative Luminescence Spectrometry in Biomedical Sciences
Ghent, Belgium: 3rd-6th September 1984

3rd International Conference on Nickel Metabolism and Toxicology
Paris, France: 4th-7th September 1984

2nd International Congress on Automation in the Clinical Laboratory
Barcelona, Spain: 15th-18th October 1984

2nd International Symposium on Harmonization of Collaborative Analytical Studies
Washington, DC, USA: 25th-27th October 1984

MINUTES OF 32nd COUNCIL MEETING

DANMARKS TEKNISKE HØJSKOLE, LYNGBY, 25–26 AUGUST 1983

Present: Prof. S. Nagakura (President, in the chair), Members of Bureau, Delegates of National Adhering Organizations, Observers from Observer Countries and Associated Organizations, Presidents of National Chemical Societies

1/83 INTRODUCTORY REMARKS AND FINALIZATION OF AGENDA

The Council stood in silence for a short period in remembrance of the colleagues deceased since the last General Assembly [details in meeting papers plus Prof. S. Mizushima (3rd August 1983) - Bureau (1955-1961)].

At the suggestion of the President, Council agreed to consider Item 9 before Item 8 and the Agenda for future meetings should be prepared accordingly.

2/83 APPROVAL OF MINUTES OF 31st COUNCIL MEETING

The Minutes of the previous Council meeting, as circulated to NAOs (82.07.23), were approved unanimously by the National Delegates.

3/83 ANNOUNCEMENT OF NOMINATIONS FOR OFFICERS AND BUREAU MEMBERS

The papers distributed before the meeting included the nominations received by the statutory deadline (83.06.25), together with biographical notes on each candidate, for the Offices of Vice-President, Secretary General, and Treasurer, and for vacancies amongst Elected Members of the Bureau. It was noted that Comité National Français de la Chimie had yesterday withdrawn the nomination of Mr. P. Poirier as a candidate for the Treasurership. Prof. Nagakura announced that the nomination of Dr. F. Wollrab for the same Office had today been withdrawn by Comité National Belge de Chimie.

On the recommendation of the Bureau (Minute 5.1/83, 47th Meeting - Lyngby), it was resolved unanimously that:

(i) there should be 11 Elected Members of the Bureau for the period 1983-5 (Statute 7.2);

(ii) the Secretary General, the Treasurer, and Dr. R. W. J. Keay (non-voting Secretary of the UK-Delegation) should act as tellers during all voting.

Council noted that, in accordance with Bylaw 2.222, the Bureau had made recommendations (Minute 5.1/83, 47th Meeting - Lyngby) for filling various vacancies. These recommendations were not binding on Council.

The FRG-Delegation reminded Council that at Leuven (Minute 3/81, 31st Council) it had withdrawn the nomination of Prof. H. Gg. Wagner as a candidate for Vice-President of the Union, so that Dr. W. G. Schneider would be unopposed.

4/83 ANNOUNCEMENT OF TIME OF ELECTIONS

Council accepted the recommendation of the Bureau (Minute 5.2/83, 47th Meeting - Lyngby) that the elections for Officers of the Union and for Elected Members of the Bureau should take place at 10.15 on 26th August.

5/83 STATUTORY REPORT OF PRESIDENT ON STATE OF UNION

Prof. Nagakura drew attention to efforts made over the past two years along lines set up by his predecessor, Prof. H. Zollinger. Excluding financial difficulties, the main problems confronting the Union were:

(i) To open ways for as many chemists as possible to participate in IUPAC activities - progress had been achieved towards an Affiliate Membership Scheme, on limitation of total service for an individual in IUPAC bodies, and recruiting Observer Countries.

(ii) To make the traditional activities of various IUPAC bodies more substantial and more efficient - Vice-President Schneider had successfully conducted during 1982-3 the first statutory critical assessment of programmes and projects of IUPAC bodies.

(iii) To introduce new roles which IUPAC should play in relation to human society - the scientific and financial success of CHEMRAWN II (Manila, 1982.12.06-10) in particular was noted.

(iv) To disseminate information on IUPAC activities accurately and effectively among as many chemists as possible and to deepen their interests in the Union - a change in emphasis of the contents of the news magazine Chemistry International had been started in 1983, wider dissemination of IUPAC publications at lower prices was necessary, and the relationship with national chemical societies was being improved.

6/83 BIENNIAL REPORT OF TREASURER

The report of the Treasurer and the audited accounts for 1981-2 had been circulated before the meeting. Because of a number of favourable factors there was an excess of income over expenditure of US-$91 651 for the biennium. The financial status (as at 83.12.08) showed a cash position of about US-$85 900 with a further US-$330 000 in short-term investments.

Dr. W. Graulich announced the following changes to the precirculated report:-

(i) Page 3, first paragraph, lines 5-7: amend the wording to read as follows -

"to change our investment policy to a more flexible approach. When the stock market declined, we sold nearly all shares out of our portfolio and invested in high-yielding, short-term deposits and bonds. At the same time we switched from the Deutschmark and Swiss-franc to the US-dollar. The aim was to protect the funds of the Union and to get a high return on the investments."

(ii) Page 3, fourth paragraph, line 8: amend the wording to read as follows -

"Since the decision to finish the old-style Chem. Intl."

Furthermore, the changes in respect of CI were not the only factor in reversing the publications deficit of 1981. Other significant contributions were the higher royalties received and no expenditure in 1982 on Membership Lists.

Extra points drawn to the attention of Council by the Treasurer were:

(i) All 1981 outstanding national dues were now cleared and there had been a reasonable subscriptions response so far for 1982.

(ii) Expenditure on Lyngby was more or less in line with budget, taking account of the higher than expected attendance of Titular Members.

Council resolved unanimously that the audited accounts for 1981-2 should be accepted and that Neutra Treuhand AG should be reappointed as auditor for IUPAC for the biennium 1983-4.

7/83 REPORT OF FINANCE COMMITTEE

A short report from the Chairman of the Finance Committee was tabled at the meeting. Dr. H. F. Wilson mentioned the provision ("bad debts") introduced in 1982 on account of

NAOs unable to pay their subscriptions in full under the new Leuven dues structure. He presented to Council a graph, showing changes in total IUPAC investments over recent years, and a corresponding table: peaks and troughs might be smoothed out in future by professional management of the IUPAC portfolio.

8/83 TENTATIVE BUDGETS FOR 1984 AND 1985

Dr. Graulich outlined the general purposes and publications budgets for 1984 and 1985 as revised by the Finance Committee at Lyngby (83.08.21). He drew attention to provision for a word-processing system at the IUPAC Secretariat, for possible introduction of the Affiliate Membership Scheme from the start of 1985, and for increased interest and dividends earned. A small surplus (US-$1 500) was anticipated for 1984, but a deficit (US-$70 000) for 1985. The two main sources of the overall deficit (US-$68 500) were:

(i) Affiliate Membership Scheme - US-$65 000,
(ii) bad debts - US-$155 000.

The deficit could, however, be met from IUPAC reserves: supplementary assets at the end of 1982 were US-$224 560.

The UK-delegation expressed concern at the increasing annual provision for bad debts, and suggested that the President should visit countries with significant subscription difficulties to convince the relevant authorities of the valuable work being done by the Union.

Several other NAOs were worried about the budgets. The FRG- and Swiss- Delegations felt that the Affiliate Membership Scheme would place an unduly heavy burden on the Union's finances. Prof. T. S. West pointed out that, even if the Scheme was not adopted, the burden would not necessarily be eliminated: there was still a need to replace the temporary part-time editor of CI in some way to maintain an IUPAC communications system.

At this point in the meeting Council proceeded to consider other items on the Agenda, while the Treasurer, Secretary General, Dr. Wilson, and Prof. West withdrew to formulate new proposals. Subsequently (after Item 20/83 on the following day) Prof. G. Ourisson presented two further amendments to the budgets (83.08.25):

(i) Assuming that for 1984 Council wanted a level of NAO dues income no higher than for 1983 (US-$400 000), invoices might be issued at a level of US-$430 000, but allowance made to consider requests (based on the actual capacity of an NAO to pay) for reductions to a maximum of US-$30 000 (i.e. reduce bad debts to a total of US-$45 000).

(ii) The seed money needed to start the Affiliate Membership Scheme in 1984 and to continue the work in 1985 (US-$25 000 each year) should be transferred from the IUPAC reserves. Some additional money would still be required to fund Information Officer activities (even part-time) from economies elsewhere in the budget.

There was strong opposition within Council towards any transfers from the reserves in the coming two years and the suggestion was defeated formally (58 in favour, 60 against). Council then resolved unanimously (132 in favour) to accept the tentative budgets as detailed in Appendix A.

9/83 REVISED NEW DUES STRUCTURE AND FIXING ANNUAL DUES FOR 1984 AND 1985

Dr. Wilson referred briefly to studies by the Finance Committee of alternative criteria to chemical turnover as the basis of IUPAC dues. Compared with CT, chemical publications (Chemical Abstracts) gave rise to a substantially different distribution of dues and the overall situation was badly distorted, despite the practical advantage of ease of obtaining national chemical publications totals.

As promised at Leuven (Minute 9/81, 31st Council), the Finance Committee had examined and developed the UK suggestion of a mathematical formula, which would make it easier to justify the provision of funds to meet national dues. A proposed improvement to the Leuven dues structure had been circulated (83.05.18, revised 83.07.14) by the Secretary General on behalf of the Executive Committee to NAOs. The dues for USSR in Appendix III should, however, be US-$40 000 in both 1984 and 1985 and the dues invoices for member countries would be based on latest available CT-figures.

Prof. Ourisson read out a statement from the Bureau (Minute 5.4/83, 47th Meeting - Lyngby), accepting that special circumstances may prevent some NAOs from contributing to IUPAC funds in accordance with the provisions of either the present or any revised scheme of dues structure. It was emphasized that the proposed improvement did not remove the existing possibilities whereby exceptions might be made to meet such difficulties. In addition, the Bureau recommended that Council should not vote at Lyngby on Appendix IV but, at this time, should simply accept for information the proposals to extend voting rights for different categories of membership.

The Dutch Delegation supported the principle of a progressive scale for dues, but Netherlands had difficulties with respect to CT and was dissatisfied with the proposed mathematical formula. Many NAOs had now reached the maximum which they could pay in practice. The only way forward for the Union over the next few years seemed to be by reducing the level of activity. These views were supported by Belgium which, in view of serious national recession, found it difficult to justify the need for continuing large increases in IUPAC dues, especially in comparison with other international scientific unions.

Concern was also expressed by the Delegations from France, GDR, Malaysia, Portugal, Spain, and Turkey, attention being drawn to the financial problems for smaller countries to join and participate in IUPAC. In the case of GDR, it was advised to submit an alternative CT-figure for consideration by the Category Subcommittee.

While recognizing that nearly all countries were experiencing difficulty in paying increases in IUPAC dues during times of recession, the UK-Delegation was convinced that the proposed formula provided a fairer mechanism for the allocation of dues. Because of the strengthening of the US-dollar, perhaps the formula might eventually be linked to a 'basket' of currencies. Council noted that, in response to a suggestion from Japan, the Finance Committee after Lyngby would be investigating a system with 1984-5 dues based on the biennial CT-average for 1981-2.

After some confusion in an initial voting, Council resolved (76 in favour, 46 against) to accept the proposed mathematical formula for implementation in 1984.

10/83 APPLICATION FOR MEMBERSHIP OF IUPAC

On the recommendation of the Bureau (Minute 5.5/83, 47th Meeting - Lyngby), Council resolved unanimously that the application for membership of the Union from Sociedad Química del Perú should be accepted.

11/83 APPLICATIONS FOR ASSOCIATED ORGANIZATION STATUS

On the recommendation of the Bureau (Minute 9.1/82, 46th Meeting - Prague; Minute 5.6/83, 47th Meeting - Lyngby), Council resolved unanimously that applications for AO status of the Union from the European Federation of Corrosion, International Organization for Biotechnology and Bioengineering, and International Zeolite Association, should be accepted. Council also accepted the recommendation that, notwithstanding Bylaw 5.5,

(i) activity reports should not be sought from the existing AOs in 1983,

(ii) the review of membership of each AO should be made in future every four years instead of at each General Assembly,

and it noted that subsequent to Leuven satisfactory reports had been obtained on the current activities of:

Association of Editors of European Chemistry Journals
International Association for Advancement of High Temperature Science and Technology
International Commission on Applications of the Mössbauer Effect
International Committee on Rheology
International Congress of Catalysis
International Organization of Crystal Growth
International Society of Electrochemistry
International Society of Magnetic Resonance

12/83 REPORTS OF DIVISION PRESIDENTS

Reports from the various Division Presidents had been precirculated, except for those of the Inorganic and Organic Chemistry Divisions which were distributed at the meeting.

Prof. N. Sheppard announced that 14 documents from Commissions of the Physical Chemistry Division had been approved at Lyngby and publication was expected during 1983-5. About 15 more documents were anticipated during the next biennium. Arrangements were well advanced for publication of Oxidation-Reduction Potentials in Aqueous Solution (a revision of Latimer's well known book). A new edition of Manual of Symbols and Terminology for Physicochemical Quantities and Units (the Green Book) was likely to be completed by 1986-7.

Attention was drawn by Dr. Mary Good first to two matters discussed at the meeting of Division Presidents in Lyngby: the new mechanism for publication of IUPAC recommendations on nomenclature and symbols, the trial period of which was to be extended from Lyngby to Lyon, and concern about the present level of dissemination of Commission reports. The Subcommittee for Assessment of Isotopic Composition had completed its work and been terminated. Regarding revision of Nomenclature of Inorganic Chemistry (the Red Book), Part I should be completed by 1985, when a final draft of chapters for Part II should also be available. Good progress had been made on Handbook of Thermodynamic and Transport Properties of the Alkali Metals (52 chapters in 2 volumes).

Prof. J. Mathieu indicated that with 19 approved projects, the Commission on Nomenclature of Organic Chemistry was particularly active, but care was needed not to over-extend itself. The work on physical organic chemistry (12 projects) was also at a high level, but less so on photochemistry (5 projects). He reiterated his view that IUPAC should establish an interdivisional activity on computers in chemistry.

Revised guidelines had been approved at Lyngby by the Macromolecular Division, covering the frequency and organization of IUPAC-sponsored symposia in macromolecular science (both multi- and single-topic). Prof. C. H. Bamford also reviewed progress during the Assembly by the Commissions on Macromolecular Nomenclature and on Polymer Characterization and Properties.

On behalf of the Analytical Chemistry Division, Prof. F. Pellerin announced the completion of 20 projects during the Assembly and the initiation of 30 new activities. Several contributions from the Commissions for the proposed book on Trace Elements of Metals and Metalloids in Water had been collected at Lyngby. Ten new chapters were planned for a revised edition of Compendium of Analytical Nomenclature (the Orange Book) being assembled in 1984.

Further consolidation of the activities of the Applied Chemistry Division had been achieved according to Dr. H. Frehse: of 39 existing projects 10 had been terminated and 3 deleted, and 8-10 new ones approved. There should be an improvement in the work of the Commission on Biotechnology and a future standing role was foreseen in air and water chemistry.

The most important achievements of the Clinical Chemistry Division at Lyngby had been reorganization of the Commissions on Automation and Clinical Chemical Techniques, on Quantities and Units, and on Teaching, in response to the IUPAC Vice-Presidential critical assessment. In addition, the Subcommittee on Clinical Chemical Standards had been dissolved and a Subcommittee on Toxicology of Chromium established.

Whilst acknowledging that the reports of the Divisions were very satisfactory, the USSR Delegation proposed that, at future General Assemblies, greater attention should be devoted to ensuring better geographical representation in the Divisions of good scientists from throughout the world.

On the recommendation of the Bureau (Minute 8.4/82, 46th Meeting - Prague; Minute 5.7/83, 47th Meeting - Lyngby), Council resolved unanimously that:-

(i) the following additional wording to paragraph 5, Article 4, of the Clinical Chemistry Division Rules (Minute 12(i)/81, 31st Council - Leuven) should be approved:

> "Commission Chairmen who are not members of the Division Committee shall attend its meetings as observers. National Representatives to the Division Committee may attend its meetings as observers."

(ii) the revised rules of the Applied Chemistry Division should be approved

(iii) the following changed wording for Rule 3.4 of the Organic Chemistry Division to that approved at Davos (Council Minute 12/79) should be approved:

> "The Division Committee shall normally appoint Co-opted Members of the Committee for terms of six years, not renewable. Terms of Co-opted Members shall be arranged so that two expire at the end of each General Assembly."

A listing of "Practices of the Organic Chemistry Division" (how the Division currently operates within the Rules) was noted by the Council.

13/83 REPORT OF COMMITTEE ON PUBLICATIONS

Dr. D. P. den Os presented some supplementary remarks to his precirculated report, taking into account discussions at the Publications Committee meeting and with various Divisions at Lyngby. He referred to the need for originators of IUPAC documents to be more aware of constraints to and opportunities for wider distribution of the results of their work (e.g., by application, from time to time, of the Executive Committee guideline "that at least the publication budget should be balanced"); to the effects of establishing Editorial Boards for CI and for PAC; to possibilities for commissioning reference books within the Divisions to fill gaps in the market; and to IUPAC's possible role in numeric chemical data bases.

A statement was read out by the USA-Delegation and views were expressed by the UK-Delegation about dissatisfaction with the cost of Pergamon-produced publications and the lack of dissemination of IUPAC documents. Dr. den Os stressed that, at Lyngby, the Publications Committee had already recommended, besides renegotiating with Pergamon when the current contract terminated at the end of 1984, a number of other publishing houses and not-for-profit societies should be invited to quote for the full IUPAC publications programme or for parts of it.

14/83 REPORT OF CHEMRAWN COMMITTEE

The major activity since Leuven was the successful completion of CHEMRAWN II: International Conference on Chemistry and World Food Supplies (Manila, 82.12.06-10). Dr. B. W. Rossiter tabled a summary report and copies of the recently printed Perspectives and Recommendations booklet. The Proceedings volume was expected shortly. Nearly US-$600 000 was raised in direct support of the Conference. An eventual surplus of about US-$18 000 was now anticipated and US-$15 000 had already been repaid to IUPAC.

Very satisfactory progress had been achieved in the arrangements for CHEMRAWN III (Den Haag, 84.06.25-29) and progress was escalating for CHEMRAWN IV (Woods Hole, 85.09.22-28). CHEMRAWN V was now tentatively entitled "Contributions of Chemistry to Human Health: The New Frontiers" and would be held in FRG (1986), FRG having withdrawn its original offer to host a meeting that year on "Water and Waste Water Technology". CHEMRAWN VI would be on "Special Materials to Meet New Needs in Energy, Transportation, and Communication" in Japan (1987). A 2nd World Conference on Chemistry, Agriculture, and World Food Supplies might take place during 1989/90.

15/83 LANGUAGE FOR OFFICIAL RECORDS DURING 1983-7

On the recommendation of the Bureau (Minute 5.8/83, 47th Meeting - Lyngby), Council resolved unanimously that English should be the one language in which the official records of the Council, Bureau, and Executive Committee should be kept and published during the four years 1983-7 (Statute 5.405).

16/83 PROPOSED CHANGES TO STATUTES AND BYLAWS

(i) On the recommendation of the Bureau (Minute 3.1/82, 46th Meeting - Prague), Council resolved (126 in favour, 0 against) that in order to expedite publication of the new Membership Lists of IUPAC Bodies after a General Assembly, the sentence in lines 30-32, page 13 of Bylaw 4.304 should be reworded as follows:

"If approval by the latter be obtained or if no reply to the request for approval be received within two months, the Bureau or Executive Committee shall take a decision at its next meeting."

(ii) Further to the deliberations of Council at Leuven (Minute 15(ii)/81), a recommendation was considered from the Bureau (Minute 3.1/82, 46th Meeting - Prague) that Statute 3.3 should be reworded. Although the recommendation was approved by Council (76 in favour, 44 against), no change could be made to Statute 3.3 because it did not receive the support of two-thirds of the total possible votes (Statute 14.3: 42 possible voting Delegations with 146 possible votes, 3 non-voting Delegations).

(iii) The Dutch-Delegation pointed out that acceptance of the mathematical formula for national dues (Minute 9/83) was in contradiction with dues being "in accordance with its category of membership" in Statute 9.1. A proposal (Minute 3.1/82, 46th Bureau - Prague) to reword Statute 9.1 was therefore withdrawn for reconsideration by the Bureau.

(iv) On the recommendation of the Bureau (Minute 3.1/82, 46th Meeting - Prague), Council resolved (120 in favour, 0 against) to delete the word "annually" from Bylaw 4.1065 in order to reflect current practice in the Union.

17/83 PROPOSED AFFILIATE MEMBERSHIP SCHEME

Prof. West referred briefly to the work since Leuven (Minute 16/81, 31st Council) of the Committee on Affiliate Membership, culminating in a poll conducted in five member countries at the end of 1982. On the basis of the results, the Executive Committee had recommended launching the Scheme from the start of 1985. At Lyngby the Bureau (Minute 5.9/83, 47th Meeting) had confirmed its support for the Scheme. Views on the subject within most Divisions were balanced.

The Delegations from Argentina, Australia, France, FRG, Netherlands, Sweden, Switzerland, UK, USA, and USSR participated in the ensuing debate and diverse opinions were expressed, from full support to total opposition or deferring introduction of the Scheme for two years. Some of the points made were:

(i) a need to identify additional benefits in order to attract Affiliates,

(ii) in addition to CI, IUPAC activities should be publicized through national chemical news journals,

(iii) the international scientific unions did not normally allow individual membership,

(iv) the USA-Delegation promised to underwrite the Scheme to the extent of US-$16 000 in the first year,

(v) if the Scheme was unsuccessful, it should be discontinued after three to four years.

Eventually a vote was taken and the proposal was approved (60 in favour, 44 against; simple majority 53).

The Secretary General commented that such a divided vote had clear implications for the Executive Committee, and that it should consider carefully the various remarks made in Council and the views of the Divisions. The Treasurer said that, although the Scheme had been approved, its implementation could be delayed by budgetary restrictions (Minute 8/83).

18/83 PROPOSALS FORMALLY RECEIVED FROM NATIONAL ADHERING ORGANIZATIONS

Prof. Zollinger reported on the deliberations of the committee appointed to look into the proposal for a limitation of total time in office for one person on various IUPAC bodies (Minute 17/81, 31st Council - Leuven). It had been established that the main concern of FRG was not with upward movement from a Commission to Division Committee or higher office of the Union, but with lateral movement from Commission to Commission. At Lyngby the Bureau (Minute 3.3/83, 47th Meeting) had accepted the recommendation that exceptional circumstances should be established and special permission of the Bureau obtained for membership of two or more Commissions of the same Division or of different Divisions, or of different Division Committees, beyond twelve years. The necessary amendments to Bylaw 4.304 would be brought before Council in 1985, but they had been adopted by the Bureau as guidelines for 1983-5.

In view of the satisfactory progress made in changing the emphasis of the contents of CI to concentrate on IUPAC activities, particularly issue No.4/1983, the FRG-Delegation withdrew its proposal to discontinue the news magazine at the earliest possible date.

Regarding the proposal of FRG to create "IUPAC Seminars", Council accepted the suggestion (Minute 5.10/83, 47th Bureau - Lyngby) to establish a study group to look in detail into the matter and to report back to the Bureau.

19/83 ADOPTION OF RECOMMENDATIONS ON NOMENCLATURE AND SYMBOLS

The following recommendations, amended as necessary from the provisional form and adopted since Leuven by IDCNS on behalf of Council (Minute 16/79, 29th Meeting - Warsaw), were ratified:

IUB-IUPAC Joint Commission on Biochemical Nomenclature (JCBN)

Nomenclature of Tocopherols and Related Compounds (PAC 1982, 54 (8), 1507-1510)

Physical Chemistry Division

Notation for States and Processes, Significance of the Word "Standard" in Chemical Thermodynamics, and Remarks on Commonly Tabulated Forms of Thermodynamic Functions (Commission on Thermodynamics) (PAC 1982, 54 (6), 1239-1250)

Nomenclature for Transport Phenomena in Electrolytic Systems (Commission on Electrochemistry) (PAC 1981, 53 (10), 1827-1840)

Presentation of Raman Spectra in Data Collections (Commission on Molecular Structure and Spectroscopy) (PAC 1981, 53 (10), 1879-1885)

Inorganic Chemistry Division

Nomenclature of Inorganic Chemistry: II.1 - Isotopically Modified Compounds (Commission on Nomenclature of Inorganic Chemistry) (PAC 1981, 53 (10), 1887-1900)

Nomenclature of Inorganic Chemistry: II.2 - The Nomenclature of Hydrides of Nitrogen and Derived Cations, Anions, and Ligands (Commission on Nomenclature of Inorganic Chemistry) (PAC 1982, 54 (12), 2545-2552)

Organic Chemistry Division

Revision of the Extended Hantzsch-Widman System of Nomenclature for Heteromonocycles (Commission on Nomenclature of Organic Chemistry) (PAC 1983, 55 (2), 409-416)

Glossary of Terms Used in Physical Organic Chemistry (Commission on Physical Organic Chemistry) (PAC 1983, 55 (8), 1281-1372)

Analytical Chemistry Division

Recommendations for Publishing Manuscripts on Ion-Selective Electrodes (Commission on Analytical Nomenclature) (PAC 1981, 53 (10), 1907-1912)

Recommendations for Publication of Papers on Precipitation Method of Gravimetric Analysis (Commission on Analytical Nomenclature) (PAC 1981, 53 (11), 2303-2306)

20/83 PROPOSALS FOR NEW BODIES AND DISSOLUTION OF EXISTING BODIES

Prof. Zollinger congratulated Dr. Schneider on the excellence of his statutory biennial critical assessment of the programmes and projects of IUPAC bodies. The assessment had been made available to the Bureau and to NAOs on 83.05.18, three months prior to the Lyngby Council meeting, as requested at Leuven (Minute 15(iii)/81, 31st Council).

Council resolved unanimously that:-

(i) the name "Commisson on Automation in Clinical Chemistry (VII.1) should be changed to "Commission on Automation and Clinical Chemical Techniques" (Minute 8.4/82, 46th Bureau - Prague);

(ii) the Commissions on Physicochemical Measurements and Standards (I.4) and on Reclamation of Solid Wastes (VI.7) should be dissolved (Minutes 5.11(i) and (iii)/83, 47th Bureau - Lyngby);

(iii) a Commission on Chemical Kinetics should be created, taking account of the Bureau discussions at Lyngby on membership, finance, and activities (Minute 5.11(ii)/83, 47th Meeting);

(iv) its authority should be delegated to the Bureau (Executive Committee) to receive after Lyngby written proposals on, and if found acceptable to re-establish the following Commissions "suspended" at Leuven:

Medicinal Chemistry (III.4)
Analytical Radiochemistry and Nuclear Materials (V.7)
Atmospheric Environment (VI.4)
Water Quality (VI.6)

(Minute 5.11(iii)/83, 47th Bureau - Lyngby);

(v) all other existing Commissions of the Union should be allowed to continue for the next two years (Bylaw 4.302).

21/83 RATIFICATION OF DECISIONS TAKEN BY BUREAU AND EXECUTIVE COMMITTEE SINCE 31st GENERAL ASSEMBLY

All decisions taken by the two bodies since those approved by Council at Leuven (Minute 20/81, 31st Meeting), were contained in the Minutes of the 43rd-46th Bureau and 98th-102nd Executive Committee Meetings. These Minutes had been circulated to all NAOs.

Council resolved unanimously that the decisions made by the Bureau and Executive Committee since the 31st General Assembly should be declared ratified.

22/83 IMPORTANT MATTERS DISCUSSED BY BUREAU AT 32nd GENERAL ASSEMBLY NOT COVERED BY ITEMS ON COUNCIL AGENDA

There were no important matters requiring the attention of Council.

23/83 ELECTIONS

There were 37 Delegations present with Category A3-C (inclusive) membership, constituting a total of 132 votes, and 3 Delegations with Category D membership (non-voting on non-scientific matters). Five NAOs (Austria, Brazil, Colombia, Syria, Yugoslavia) were not represented.

23.1/83 Vice-President

In a written and secret ballot:

Prof. C. N. R. Rao	72
Prof. H. Gg. Wagner	60

(total voting 132, simple majority 67)

Prof. Rao was declared the elected Vice-President.

23.2/83 Secretary General

Prof. T. S. West was declared unanimously the elected Secretary General.

23.2/83 Treasurer

In a written and secret ballot:

Prof. A. Björkman	96
Dr. F. J. M. Nieuwenhuis	36

(total voting 132, simple majority 67)

Prof. Björkman was declared the elected Treasurer.

23.4/83 Elected Members to Bureau

In a written and secret ballot for 6 vacancies:

*Prof. Y. Jeannin	124
*Prof. H. Gg. Wagner	122
*Prof. V. A. Koptyug	120
Dr. A. J. Arvía	104
*Prof. D. B. Tonks	94
*Prof. A. A. Vlček	94
Prof. S. Itô	74
Prof. R. L. Martin	56

(* re-election for a further period of four years;
total voting 792, simple majority 67)

The 6 vacancies were declared to be filled by Dr. Arvía and Profs. Jeannin, Koptyug, Tonks, Vlček and Wagner.

The Australian-Delegation expressed concern at the practice of the Bureau, in making recommendations for filling vacancies on the Bureau, to give priority to candidates offering themselves for a second period of four years. That reduced the possibilities for election of new candidates. Accordingly, the Bureau was requested to report to Council in 1985 the principles which it had adopted in selecting its recommended list from the nominations for Elected Members, especially with respect to fair geographical distribution of Members of the Bureau (Statute 7.2).

23.5/83 Officers of Divisions

The following elections of Officers were approved unanimously by Council (Bylaw 4.104):

Physical Chemistry Division

President	(1983-1987)	Dr. D. R. Lide, Jr.
Vice-President	(1983-1987)	Prof. K. I. Zamaraev
Secretary	(1981-1985)	Dr. E. D. Becker

Inorganic Chemistry Division

President	(1981-1985)	Dr. Mary L. Good
Vice-President	(1981-1985)	Prof. Kazuo Saito
Secretary	(1981-1985)	Prof. E. Fluck

Organic Chemistry Division

President	(1983-1985)	Prof. G. Modena
Vice-President	(1983-1985)	Prof. J. F. Bunnett
Secretary	(1983-1987)	Dr. Sukh Dev

Macromolecular Division

President	(1981-1985)	Prof. C. H. Bamford
Vice-President	(1981-1985)	Prof. T. Saegusa
Secretary	(1981-1985)	Prof. M. Mandel

Analytical Chemistry Division

President	(1981-1985)	Prof. F. Pellerin
Vice-President	(1983-1985)	Prof. G. H. Nancollas
Secretary	(1983-1987)	Prof. G. den Boef

Applied Chemistry Division

President	(1981-1985)	Dr. H. Frehse
Vice-President	(1981-1985)	Dr. P. C. Kearney
Secretary	(1981-1985)	Dr. L. E. Coles

Clinical Chemistry Division

President	(1981-1985)	Prof. F. W. Sunderman, Jr.
Vice-President	(1981-1985)	Dr. S. S. Brown
Secretary	(1981-1985)	Prof. N. Montalbetti

24/83 RATIFICATION OF DATES AND PLACES OF 33rd GENERAL ASSEMBLY AND 30th CONGRESS

24.1/83 33rd General Assembly

On the recommendation of the Bureau (Minute 5.2/82, 46th Meeting - Prague), Council ratified Lyon-Écully, France, as the location and 30th August-7th September 1985 as the dates for the 33rd IUPAC General Assembly.

24.2/83 30th Congress

Council ratified its acceptance (Minute 24.2/81, 31st Meeting - Leuven) of the invitation from UK to host the 30th IUPAC Congress, with Manchester as the location and 9th-13th September 1985 as the dates. Also, it approved the scope of the scientific programme (Statute 13.1) as indicated in the preliminary announcement, but subject to any suggestions for changes by the Division Presidents and clarification of the use of "Division" for the next circular (Minute 5.12/83, 47th Bureau - Lyngby).

25/83 PROVISIONAL PLACES OF 34th GENERAL ASSEMBLY AND 31st CONGRESS

25.1/83 34th General Assembly

On the recommendation of the Bureau (Minute 5.13(i)/83, 47th Meeting - Lyngby), Council accepted, in principle, an invitation from USA to host the 34th IUPAC General Assembly at Boston University during 21st-29th August 1987.

25.2/83 31st Congress

On the recommendation of the Bureau (Minute 5.13(ii)/83, 47th Meeting - Lyngby), Council accepted, in principle, an invitation from Bulgaria to host the 31st IUPAC Congress at Sofia in 1987.

25.3/83 35th General Assembly and 32nd Congress

Council noted that Sweden had issued a definite invitation to host the 1989 Assembly at Lund and would discuss later this year whether it might also host the 1989 Congress in Stockholm (Minute 15(i)/83, 47th Bureau - Lyngby).

26/83 ANY OTHER BUSINESS

There being no other business, Prof. Nagakura thanked Prof. A. Kjaer and his colleagues for the excellent local organization and hospitality. After handing over the Presidency to Dr. Schneider, the latter paid tribute to the retiring Officers of the Union and two retiring Division Presidents (Profs. Sheppard and Mathieu). He was distressed at the financial situation of IUPAC and the inability to fund fully the activities of the Divisions, but promised, with his colleagues in the Executive Committee and Bureau, to address himself to these important problems.

APPENDIX A

IUPAC GENERAL PURPOSES BUDGET 1984 *

INCOME

Formal dues from National Adhering Organizations	400 000
Company Associates Service Costs	5 000
Interests and Dividends Earned	60 000
Subvention from ICSU	15 000
Transfer from Publications Account	9 500
	489 500

EXPENDITURE

Office Expenses		
Secretary General & Officers	3 000	
IUPAC Secretariat	147 000 +	
Audit, Bank Fees & Other Charges	20 000	170 000
Travel, Subsistence, Administration of IUPAC Bodies		200 000
Contributions to Symposia		5 000
Other Expenses		
Contributions to ICSU & ICSU Committees	9 500	
IUPAC Representatives on Other Organizations	7 500	
Transfer to Publications Account for Free Chem. Intl.	16 000	33 000
Provison for Bad Debts		45 000
General Contingency Fund		10 000
		463 000
Excess of Income over Expenditure		26 500
		489 500

* All figures in US-dollars
\+ Includes provision for word-processing system

IUPAC PUBLICATIONS BUDGET 1984*

INCOME		EXPENDITURE	
Publications (Pergamon)	135 000	Scientific Editor	4 500
Publications (Secretariat) & Miscellaneous Royalties	1 000	Sci. Pubs. Sec., Ed. Chem. Intl. & Typists	74 000
		Purchase of Chem. Intl.	16 000
Transfer from General Purposes Account for Free Chem. Intl.	16 000	Share of Secretariat Costs	48 000
			142 500
		Excess of Income over Expenditure	9 500
	152 000		152 000

* All figures in US-dollars

IUPAC GENERAL PURPOSES BUDGET 1985 *

INCOME

Formal dues from National Adhering Organizations	430 000
Company Associates Service Costs	5 000
Interests and Dividends Earned	65 000
Subvention from ICSU	15 000
Transfer from Publications Account	7 500
	522 500
Excess of Expenditure over Income	42 500
	565 000

EXPENDITURE

Office Expenses		
Secretary General & Officers	4 000	
IUPAC Secretariat	155 000 +	
Audit, Bank Fees & Other Charges	20 000	179 000
Travel, Subsistence, Administration of IUPAC Bodies		
Meetings other than General Assembly	22 000	
General Assembly	270 000	292 000
Other Expenses		
Project Programmes Listing	4 000	
Transfer to Publications Account for Free Membership Lists (1983-5) & Chem. Intl.	22 000	
Contributions to ICSU & ICSU Committees	10 000	
IUPAC Representatives on Other Organizations	8 000	44 000
Provison for Bad Debts		40 000
General Contingency Fund		10 000
		565 000

* All figures in US-dollars
+ Includes provision for word-processing system

IUPAC PUBLICATIONS BUDGET 1985*

INCOME		EXPENDITURE	
Publications (Pergamon)	140 000	Scientific Editor	4 500
Publications (Secretariat) & Miscellaneous Royalties	1 000	Sci. Pubs. Sec., Ed. Chem. Intl. & Typists	78 000
Transfer from General Purposes Account for Free Membership Lists & Chem. Intl.	22 000	Purchase of Membership Lists (1983-5) & Chem. Intl.	22 000
		Share of IUPAC Secretariat Costs	51 000
			155 500
		Excess of Income over Expenditure	7 500
	163 000		163 000

* All figures in US-dollars

Printed in Dunstable, United Kingdom

91338430R00118